Microbiology of extreme environments

Environmental Biotechnology Series Editors:

Other Books in Series:

Venetia A. Saunders & Jon R. Saunders:
MICROBIAL GENETICS APPLIED TO BIOTECHNOLOGY: PRINCIPLES & TECHNIQUES OF GENE TRANSFER & MANIPULATION

Stuart B. Levy & Robert V. Miller:
GENE TRANSFER IN THE ENVIRONMENT

Henry L. Ehrlich & Corale L. Brierley:
MICROBIAL MINERAL RECOVERY

David P. Labeda:
ISOLATION OF BIOTECHNOLOGICAL ORGANISMS FROM NATURE

Forthcoming:

Morris A. Levin & Harlee S. Strauss:
RISK ASSESSMENT IN GENETIC ENGINEERING: ENVIRONMENTAL RELEASE OF ORGANISMS

J. Gregory Zeikus & Eric A. Johnson:
MIXED CULTURES IN BIOTECHNOLOGY

C.A. Reddy:
BIOCONVERSION OF PLANT BIOMASS

Microbiology of extreme environments

Edited by

CLIVE EDWARDS

McGraw-Hill Publishing Company

New York St. Louis San Francisco Auckland Bogotá
Caracas Hamburg Lisbon London Madrid Mexico
Milan Montreal New Delhi Oklahoma City
Paris San Juan São Paulo Singapore
Sydney Tokyo Toronto

Library of Congress Cataloging-in-Publication Data

Microbiology of extreme environments / edited by Clive Edwards.
p. cm.
Includes bibliographical references.
ISBN 0-07-019443-2: $42.95
1. Microbial ecology. I. Edwards, Clive, 1949–
QR100.M53 1990
576′.15—dc20
89-25528 CIP

1234567890 8965432109

ISBN 0-07-019443-2

First published in 1990 by Open University Press,
Celtic Court, Buckingham, MK18 1XT.

Typeset by Vision Typesetting, Manchester, Great Britain.
Printed in Great Britain by Alden Press, Oxford.

Contents

Contributors

Dr C. Edwards, Department of Genetics and Microbiology, University of Liverpool

Dr J.C. Fry, School of Pure and Applied Biology, University of Wales College of Cardiff

Dr G.M. Gadd, Department of Biological Sciences, University of Dundee

Dr D. Gilmour, Department of Microbiology, University of Sheffield

Dr W.J. Ingledew, Department of Biochemistry and Microbiology, University of St Andrews

Professor D.H. Jennings, Department of Genetics and Microbiology, University of Liverpool

Dr R.G. Kroll, AFRC Institute of Food Research, Reading

Introduction

Microbiology has contributed a great deal to our understanding of living processes. This has largely stemmed from work on industrially or medically important species, and the majority of these are aerobic, mesophilic, neutrophilic heterotrophs. However, as mechanisms and models for cellular processes have been established using such species, so unusual examples that contradict them are continually being discovered. Also, as novel microbial metabolites or capabilities are avidly sought in the fields of medicine and biotechnology, so attention is increasingly directed to unusual microbial species. These in turn challenge our preconceptions regarding cell biology and biochemistry. Nowhere is this more true than in those microbes that are capable of growth at environmental extremes or in conditions that were previously thought to be too harsh to support any form of life.

This book deals with microorganisms that are able to survive and grow in habitats that are considered to be hostile to most living organisms. Recently the popular all embracing term 'extremophiles' has been coined for these organisms. What constitutes an extreme environment? It is difficult to define precisely; for example, as oxygenic photosynthesis developed during the course of evolution, the anaerobic atmosphere gradually became rich in oxygen, imposing an extreme environment on the pre-existing species. Indeed oxygen, now so essential to higher life forms, could at that time be viewed as the Earth's first major pollutant! Therefore, in answer to the initial question, it rather depends on what is considered the norm. A realistic definition of an extreme habitat would be one in which species diversity is restricted to a relatively few species that have adapted in some way to the rigours of their surroundings. Such a definition must suffice despite obvious contradictions. For example, it is possible to isolate thermophilic or alkalophilic bacteria from common garden soils.

How are organisms from extreme environments adapted? Again, this will depend on the extreme condition in question. Mechanisms for adaptation might include exclusion, detoxification or adaptation. In fact, all three may be encountered depending on which group of organisms is being considered. Thermophiles, some halophiles, osmophiles and metal-tolerant species have adapted their physiology to differing extents in order to survive the harsh conditions in which they find themselves. Other metal-tolerant species have developed mechanisms for detoxifying the effects of heavy metals, particularly mercury. The acidophiles and alkalophiles survive extremes of pH largely by exclusion mechanisms, although it is probably true to say that they must also possess modified cell structures, particularly those such as the exterior-facing side of membranes or peptidoglycan which are in direct contact with the low or high pH environment.

The overall aim of the book is to provide an introduction for students at the undergraduate/postgraduate level to diverse microorganisms that inhabit extreme and unusual environments. All the contributors have attempted to include both eukaryotic as well as prokaryotic examples. In order to ensure uniformity of presentation, an attempt has been made to adhere to a fairly rigid format for each chapter. After a brief introduction, each chapter discusses physical and chemical features of the extreme environment and then goes on to introduce the types of microorganisms that are found there. The physiological and molecular basis of tolerance to the environmental conditions are then discussed, followed by a review of how these special organisms may be or are being exploited for biotechnological processes. The final major section is concerned with recent advances in the molecular genetics of the relevant extremophile. However, as will be immediately apparent to the reader, there is no uniformity in the relative weighting given to these sections between successive chapters. This illustrates our ignorance of many features of these important groups of microorganisms rather than deliberate omissions on the part of the authors.

A conscious decision has been made to deal only with environmental extremes that have been well studied. Thus omissions include those species that will grow only at low temperatures — the *psychrophiles* — and those that grow at high pressure — the *barophiles*. The main reason for this is that there is insufficient all-round knowledge of these groups to fit adequately into the format of this book. Sections which have been deemed important are the physiological and biochemical basis for adaptation to an extreme environment, biotechnological applications and the use of genetic techniques to exploit these or to unravel the molecular mechanisms that may be important to survival in a given extreme environment. Indeed, it may be justifiably said that such information is extremely limited in some of the examples that are discussed here. Notable areas include the lack of molecular genetic techniques used for studying acidophiles, halophiles or the osmophiles, and, in particular, the very limited knowledge of all aspects of oligotrophs. This last group may become increasingly studied as the pollution of drinking waters with trace organic molecules becomes a problem of some significance and urgency in the future.

Much that is discussed in this book is also a signpost to the future as well as what is possible today. In doing so all the authors have attempted to provide a balanced view of the potential of the microorganisms that are discussed since very often there is overstatement in the literature regarding their possible applications. Therefore the disadvantages of using 'extremophiles' are also stressed. We hope that anyone reading these chapters will enjoy and appreciate the unusual organisms that are described. There is no doubt that they constitute an untapped reservoir of potential benefit for a number of purposes. This book aims to summarize the current state of the art regarding these microorganisms and to attempt to herald their potential in the hope that some research workers will be stimulated to achieve its realization.

CHAPTER 1

Thermophiles

C. Edwards

Introduction

The large numbers of different genera of microorganisms now recognized serve to emphasize the diversity that is encountered in the microbial world. Diversity is also manifest within a genus and in some cases at the level of an individual species. This reflects the ability of microorganisms to adapt to a multiplicity of conditions and helps to explain how they have been able to colonize diverse environments. Temperature is a good parameter for illustrating both diversity and adaptation. No living system is indifferent to it but there are numerous ways in which organisms have adapted to respond to changes in temperature. This chapter deals with those microorganisms that are able to survive and grow at temperatures that are considered above those normally associated with biological systems. These range from those barely tolerable to the human hand to that of boiling water or even higher. By far the majority of species are bacterial and in general thermophilic eukaryotes which rarely grow at temperatures higher than 55 °C are not well described. Therefore bacterial species will be discussed mainly but, where possible, mention will be made of any relevant eukaryotic species. Sufficient information regarding the activities of thermophiles has already emerged to indicate that they have the potential for many biotechnological applications. They pose fundamental questions regarding the biochemical and genetical basis of life at high temperatures that in some aspects challenges our knowledge of the better understood mesophilic species.

WHAT DEFINES A THERMOPHILE?

Temperature ranges for growth have often been used as a means to classify groups of microorganisms. The most common divisions are the *psychrophiles*, (-3 to 20 °C);

mesophiles, (13–45 °C) and *thermophiles* (42–100 °C or more). Within each of these, subdivisions have been added in order to reconcile those species that do not fall easily into the major grouping. For example *Bacillus coagulans* has a temperature range for growth of 30–60 °C and falls into both mesophilic and thermophilic divisions. This bacterium is therefore often referred to as a *facultative thermophile*. In a sense it is more relevant to think of growth temperatures as a continuum whereby different species can be identified with optimum ranges for growth within a span from approximately 0 to 100 °C.[17] What is noteworthy is that most species appear to be restricted to a range of approximately 30–40 °C, for example, *Escherichia coli* (10–42 °C), *Bacillus stearothermophilus* (35–77 °C) and *Pyrodictium occultum* (80–110 °C). Due to the varying and in some cases overlapping responses of microorganisms to temperature, it has not been easy to form a consensus as to the definition of a thermophile. Indeed it would appear that it is impossible to arrive at a universal definition. For the purposes of this chapter any microorganism capable of growth at temperatures in excess of 55 °C will be considered as thermophilic. This will be near the limits for growth of many species, notably eukaryotes such as fungi and bacterial genera such as the *Streptomyces*.

Habitats

Natural environments that are thermophilic are widespread throughout the globe.[16,17] They are chiefly formed as a result of volcanic activity or movement of the Earth's crust at tectonically active sites. These geological phenomena result in the upward mobilization of heated water into which a range of minerals dissolve. This water forms hot springs or issues under pressure as geysers and is usually of neutral pH. Because of the extensive solubilization of such salts as chlorides, sulphates, sulphides, bicarbonates, etc. an additional physiological stress may be introduced. For example, sulphur or iron-rich hot springs have a pH of 1–2 which restricts microbial survival to thermoacidophiles such as *Sulfolobus acidocaldarius*. Deep-sea geothermal vents provide another example of highly mineralized and thermophilic environments. In these specialized habitats conditions include salinity, which increases mineralization, and pressure that allows higher temperatures. The organisms isolated from these environments are highly adapted and difficult to isolate and cultivate in the laboratory. One such bacterium isolated from a deep-sea vent is *Methanococcus jannaschii*. Thermophilic halophiles have also been isolated. These include a *Thermus* species that grew in the presence of NaCl at 3% or higher[49] and a species that was nutritionally different from *Thermus* termed *Rhodothermus marinus* that grows optimally at 65 °C pH 7 and 2% NaCl.[2]

It is perhaps not surprising that thermophilic species can be isolated from geothermal locations. However, as one of the first reports [64] of the isolation of a bacterium, that could grow at 73 °C, from the river Seine in 1888 shows, thermophiles can be isolated from quite unexpected habitats such as garden soils (e.g. *B. stearothermophilus*) and even from glaciers. Another good source of thermophilic bacteria is rotting plant matter found in garden or commercially based mushroom compost. This process comprises both mesophilic and thermophilic phases in which Gram-positive bacteria predominate. Temperatures between ambient and 80 °C may be encountered and the diverse microflora includes thermophilic bacilli, streptomycetes and other actinomycetes.[4]

The organisms

The origin and status of thermophilic microorganisms has long been a subject of some debate. In 1927 Arrhenius suggested that they originated from the planet Venus and travelled to the Earth by radiation pressure from the sun.[7] However, as Table 1.1 clearly shows, thermophily is a general phenomenon that is not restricted to a few specialized genera. The table also lists terminologies that will be used to describe the different ranges of thermophilic temperatures. These temperatures are for *growth* and not for *survival*. Many species are able to survive at higher temperatures than those that allow growth, for example, by adopting metabolic shutdown states either as vegetative cells or heat resistant spores and cysts. Photosynthetic, aerobic and anaerobic species are represented. Where the organisms are isolated from specialized thermophilic habitats, such as volcanic springs, additional extreme stresses may also be found such as low pH, high pressures or limited energy sources that require specialized energy-conserving reactions. The numbers of species decrease on moving up the temperature range. We should also note that not all thermophiles have an obvious potential for commercial exploitation. In general, as the temperature increases so do anaerobic species. This is probably due to the decreased solubility of oxygen with increasing temperature.

B. subtilis is included as a thermotolerant species since some strains are capable of growth up to 50 °C. Additionally, in one study thermophilic mutants of *B. subtilis* and *B. pumilus* have been obtained by plating out large numbers of cells followed by incubation at a temperature 10 °C in excess of the maximum for growth. Mutants obtained in this way grew in the range 50–70 °C.[26]

The *facultative thermophiles* form a large and disparate group and include any thermophilic eukaryotic species. Taxonomic studies that summarize a number of morphological and biochemical characters have recently been reported for thermophilic streptomycetes[35] and yeasts.[5]

Obligate thermophiles are represented by a range of species. The best studied are the *Bacillus*[70] and the *Thermus* group. The latter are aerobic Gram-negative yellow-pigmented heterotrophs that taxonomically show some dependence on their geographic locations.[42,65] Although the ability to fix nitrogen is widespread amongst the phototrophs, thermophilic species rarely do so. For example, nitrogen fixation could not be detected in *Chloroflexus aurantiacus*.[37] A notable exception is *Mastigocladus laminosus* which has been shown to fix atmospheric nitrogen above 50 °C.

At temperatures above 80 °C the archaebacteria predominate. There are two main phylogenetic branches in this group:

Table 1.1 Examples of the different groups of thermophilic microorganisms capable of growth at a range of high temperatures (taken largely from refs 5, 17 and 88)

	Temperature range (°C) for growth	*Comments*
Thermotolerant		
Bacillus licheniformis	20–50	Amylase active up to 90 °C
Bacillus subtilis		Well-studied genetic systems. Suitable host for cloning genes from other thermophilic bacilli
Facultative thermophiles		
Bacillus coagulans	30–60	Produces L-lactic acid
Streptomyces thermoviolaceus		Antibiotic produced up to 55 °C
Kluyveromyces marxianus	25–50	Yeast known to ferment at 48 °C
Torula thermophila		Yeast
Aspergillus fumigatus		Fungus from compost
Melanocarpus albomyces		Fungus isolated from soil or compost hydrolyses xylan
Obligate thermophiles		
Bacillus stearothermophilus	40–80	Spores exceptionally thermostable
Bacillus acidocaldarius		Thermoacidophile
Thermus aquaticus	45–79	Taq 1 polymerase used in gene amplification
Thermomonospora chromogena	37–65	Actinomycete, common and active in composts
Mastigocladus laminosus	35–64	Cyanobacteria
Synechococcus lividus	55–74	Cyanobacteria
Methanobacterium thermoautotrophicum	45–75	Methanogen
Clostridium thermohydrosulfuricum	40–78	Anaerobe
Clostridium thermocellum	40–68	Anaerobic cellulose degradation
Thermoanaerobium ethanolicus	35–78	Anaerobic, ethanol producer
Caldoactive		
Sulfolobus acidocaldarius	50–90	Thermoacidophilic S oxidizer potential application for metal leaching
Thermothrix thioparus	55–85	S oxidizer
Desulfovibrio thermophilus	50–85	SO_4 reducer
Methanococcus jannaschii	50–95	Methanogen, uses H_2 and CO_2 only
Barothermophiles		
Pyrodictium brockii	80–110	From deep-sea vents. Anaerobic autotroph

1. The strictly anaerobic methanogens together with the aerobic halophiles. To date only *Methanococcus jannaschii* and *Methanothermus fervidus* are extremely thermophilic. Most species are mesophilic.
2. The second group consists of the aerobic and anaerobic sulphur-metabolizing *Sulfolobales*, *Thermoproteales* and *Thermoplasmales*, the members of which are nearly all extreme thermophiles and to date no mesophilic species have been isolated.

CULTIVATION

Growth of thermophiles poses a number of problems that require modifications of traditional methods.[87] Cultivation of caldoactive and barothermophiles (Table 1.1) will require the development of new methods and equipment. At higher incubation temperatures there is more evaporation of both solid and liquid media. This is increased if growth in liquid culture is accompanied by sparging with gases. Condensation can lead to confluent growth on solid media which may lead to problems in isolating pure cultures from thermophilic environments or during genetic studies. Gas solubility generally decreases with rising temperature. For example, at 30 °C oxygen solubility of water is 237 nmoles ml^{-1}, at 50 °C it is 171 and at 70 °C only 120. Therefore cultivation of cultures that require gassing (aerobes or anaerobes) must ensure that there is a large liquid surface to gas ratio which can be achieved by using baffled flasks, high rates of agitation and reduction of the liquid to flask volume ratio.[87] Gas transfer rates can be improved by pressurizing the growth vessel to between 1 and 3 atm. This will be essential at temperatures in excess of 100 °C in order to keep water in the liquid state. At temperatures in excess of 65–70 °C most plastic agar dishes will melt and agar itself becomes unstable. Media constituents become more thermolabile, for example, there will be much more rapid caramelization of sugars with increasing temperature. At temperatures above 55 °C it is preferable to use agar at 2–4% to decrease evaporation. At temperatures in excess of 70 °C it is necessary to employ other solid media such as silica gel. A recent development is a clarified gum from *Pseudomonas* spp. sold under the trade name of 'Gelrite' which relies on divalent cations such as magnesium or calcium to solidify it and which is suitable as a solid medium at high temperatures. It has been used to culture extremely thermophilic microorganisms from submarine hydrothermal vents and remained solid at 120 °C and at vapour pressures and hydrostatic pressures to 265 atm.[25] For the caldoactive and barothermophilic species specialized apparatus is required for their cultivation. For example, growth in water baths is possible up to around 70 °C. However, at higher temperatures evaporation becomes a serious problem which can be overcome by using liquids that have a higher boiling point than water, e.g. glycerol. There is no doubt that if these extreme thermophiles can be demonstrated to have useful commercial properties the technology to grow them will develop apace. A good example of where this has already happened is in the study of anaerobes which is now widespread due to increased knowledge of their biochemistry and physiology which stems from improved equipment for their cultivation and maintenance.

Thermotolerance

LIMITS FOR GROWTH

The highest temperatures at which bacteria can grow and/or survive is the subject of some controversy. Currently it is thought that as long as water remains liquid, which at temperatures in excess of 100 °C requires pressures above atmospheric, then it may be possible to detect microbial life. To date unequivocal data that show growth above 100 °C have been obtained for *Pyrodictium* spp. which has an optimum growth temperature of 105 °C and a maximum in excess of 110 °C. Reports of bacteria growing at 250 °C and 265 atm. in water samples from a deep-sea vent have not been reproduced by other workers.[12] Indeed the limits to growth will be imposed by the thermostability of key cellular metabolites and it is significant that ATP, amino acids and peptides are unlikely to be thermostable at temperatures in excess of 150 °C.

How are thermophiles able to survive and flourish at temperatures up to and in excess of boiling water? There is no simple answer to this question but some progress has been made in identifying factors that stabilize macromolecules against the effects of temperature. Very often this has been achieved by comparison of the structure of a macromolecule from a thermophile with that of its mesophilic counterpart.[17,57]

THE EUBACTERIAL CYTOPLASMIC MEMBRANE

The eubacterial cytoplasmic membrane is composed of a phospholipid bilayer in which proteins are embedded. It allows the selective transfer of hydrophilic molecules, in both directions, across what is effectively a hydrophobic barrier. In many bacteria energy-conserving reactions are also located within the membrane and therefore it not only plays an important part in regulating cellular activity but in doing so also indirectly controls its own activity. Because changes in temperature can grossly affect the functioning of membrane activity, it is hardly surprising that it has long been recognized as being especially vulnerable to temperature fluctuations. Many membrane associated enzymes are in intimate contact with lipid which very often is essential for their functioning. Lipid has an important role in regulating membrane activity and its composition is particularly responsive to changes in temperature in all microbial species. This adaptive response involves a thermotrophic gel–liquid crystalline transition as temperature is increased. Normally pure liquids have a characteristic temperature for this transition that is dependent on the lipid structure. Membranes have more complex mixtures of lipids and therefore exhibit much broader transition phases especially when they also have a limited capacity for changing lipid composition with temperature. This capacity is thought to span a similar temperature range to that over which growth can occur and is in the order of 30–40 °C.[72] The gel state is reflected by a rigid membrane that allows little molecular movement. At the other extreme the liquid crystalline phase is associated with a high degree of disorder and molecular movement which, if increased above the cell's capacity to control it, will result in severe disorganization and breakdown of the membrane's ability to maintain ionic gradients and control nutrient uptake. The fluidity of the membrane is largely controlled by its fatty acid composition. The trends that have been observed[73] in mesophilic species on raising the temperature involve: 1. increased chain length of fatty acyl chains; 2. a decrease in the degree of unsaturation and 3. an increase in the proportion of methyl branched chains. Figure 1.1 (a–d) illustrates the trends observed in fatty acyl composition on decreasing the temperature. Normal straight chains have the highest melting points (Fig. 1.1 a) followed by *iso*-branched (Fig. 1.1 b), *ante iso* branched (Fig. 1.1 c) and the lowest melting points are obtained by unsaturated chains (Fig. 1.1 d).

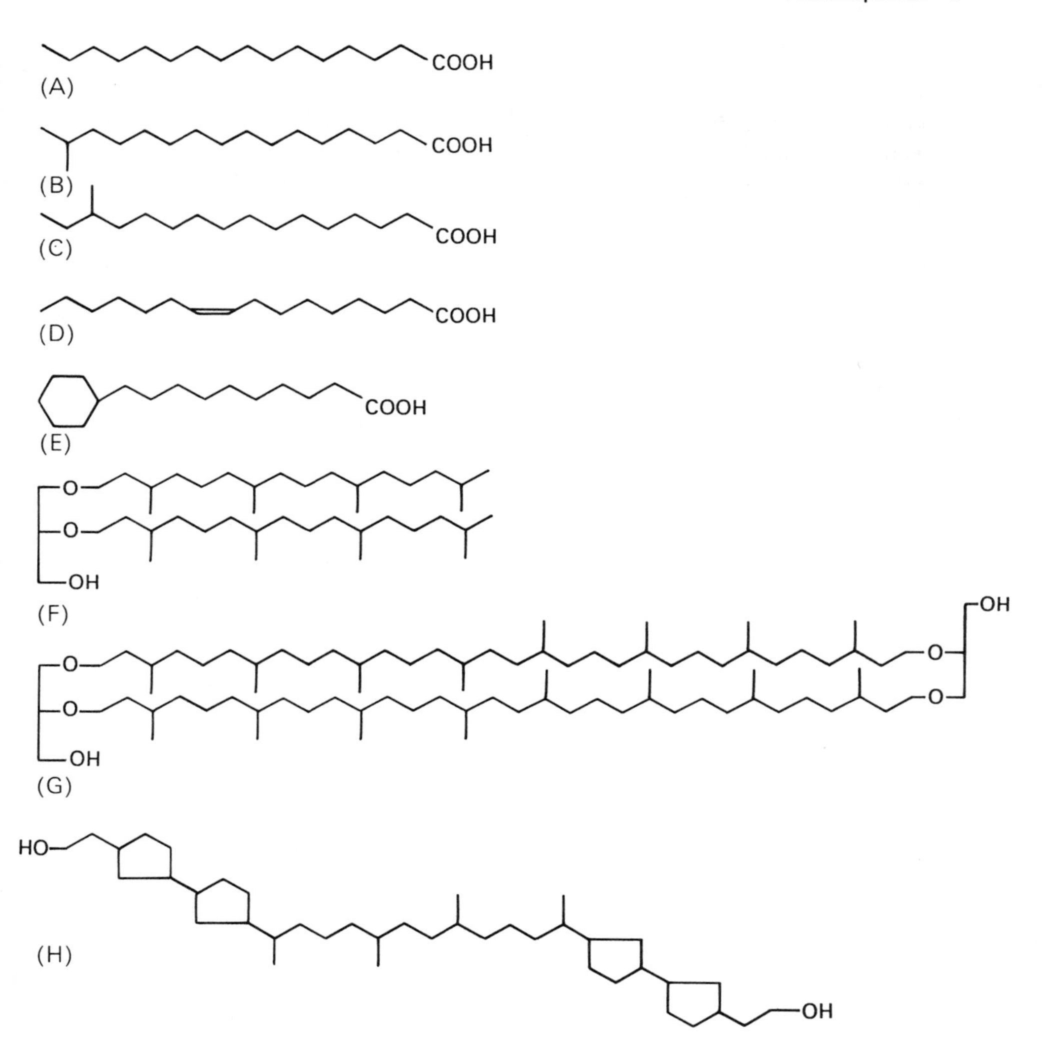

Fig 1.1 Some of the types of lipid acyl chains found in bacterial lipids. A, saturated straight chain; B, *iso*-branched; C, *ante iso* branched; D, unsaturated chain; E, alicyclic ω-cyclohexyl fatty acid; F, phytanyl glycerol diether; G, dibiphytanyl diglycerol tetraether; H, C_{40} biphytanyl diol containing four pentacyclic rings. Adapted from references 55 and 73.

LIPID COMPOSITION IN MEMBRANES OF THERMOPHILES

The type of lipid found in thermophilic bacteria and its role in thermotolerance has been recently reviewed by Langworthy and Pond.[55] Only the major trends will be discussed here.

Eubacteria

From the considerations described above it might be expected that the cytoplasmic membranes from thermophilic eubacteria consist largely of long chain normal and/or *iso*-branched fatty acyl chains. The best studied species are the *Bacillus*, especially *B. stearothermophilus*, *Thermus* and, amongst the anaerobes, *Clostridium*. In general *iso* and *ante iso* fatty acyl chains, predominate. For *B. stearothermophilus* between 34 and 64% of cell lipid consisted of *iso*-C15 and *iso*-C17 fatty acyl chains but there was also significant amounts of *ante iso*-C16 and C17 detected. In the more extreme thermophiles *Bacillus caldotenax* and *B. caldolyticus iso*-C15, C16, and C17 were found to comprise around 80% of total fatty acids. Temperature shift experiments from 45 to 85 °C resulted in shifts from *iso*-C15 to *iso*-C17 and *iso*-C16 to *n*C16. In *Thermus*, which is more thermophilic than the bacilli described above, the most abundant fatty acid was found to be *iso*-C17 (approximately 55% of total) followed by *iso*-C15. Finally, in the anaerobe *Clostridium thermohydrosulfuricum iso*-C17 (around 75% of total) and *iso*-C15 (19%) were the predominant fatty acids. An unusual fatty acid profile was seen in *C. thermocellum* which consisted largely of *iso*-C16, *n*C16 and *ante iso*-C17 but under conditions in which ethanol was an end-product of fermentation some unsaturated C14 and C16 fatty acyl chains could be detected. These were probably synthesized as a result of ethanol production and resulted in a membrane better adapted to low temperatures which adversely affected the thermotolerance of this thermophile.

An unusual component of the membrane lipids of *B. acidocaldarius* is an alicyclic ω-cyclohexyl fatty acid (see Fig. 1.1 e). This bacterium must cope with extremes of both pH and temperature and the curent evidence suggests that this unusual fatty acid has more to do with maintenance of barrier functions at low pH rather than maintaining membrane fluidity at high temperature.

Archaebacteria

Well-studied archaebacterial species include *Thermoplasma*, especially *T. acidophila*, and *Sulfolobus*, and, amongst the anaerobes, *Methanobacterium thermoautotrophicum* and *M. fervidus*. Archaebacteria are characterized by having isopranoid alcohols as the major lipid component of their membranes. Unlike the eubacteria that can adapt to temperature changes by altering chain length and degree of saturation of their fatty acyl chains, the apolar chains of archaebacteria are fixed at either twenty or forty carbon atoms. The C20 isopranoid alcohol is a fully saturated phytanol whereas the C40 isopranoids are equal to two phytanol molecules that are linked together. Attachment of these isopranoids to glycerol molecules yields a glycerol ether of which there are two types in archaebacteria. One is a phytanyl glycerol diether (Fig. 1.1f), the other is a dibiphytanyl diglycerol tetraether (Fig. 1.1g). Tetra- and diethers occur in varying ratios in both mesophilic and thermophilic archaebacteria. In the thermoacidophiles (e.g. *Thermoplasma*, *Sulfolobus*) the tetraethers predominate. The acyclic nature of the C40 chains can be altered by the inclusion of one to four cyclopentane rings. An example of a chain containing four such rings is shown in Fig. 1.1h.

The diethers are able to form a lipid bilayer in the membrane whereas the tetraethers are able to provide a monolayer, especially when their linearity is altered by the inclusion of cyclopentane rings within the molecule. Experiments in which the di- and tetraethers have been incorporated into liposomes indicate that they are able to maintain a liquid–crystalline state over an extremely wide temperature range. Tetraethers are able to form covalently linked bilayers that are able to maintain their integrity even at exceptionally high temperatures. Thus far few eubacteria have been found that can grow at temperatures above 85 °C and this may be due to an inability to maintain a lipid bilayer at these high temperatures. In the archaebacteria the covalently condensed bilayer is much stabler and does not melt apart. The highest temperature for maintenance of the integrity of archaebacterial lipids is not at present known but is likely to approach the upper limits for life. Many of these unusual lipids remain to be found, the biotechnological potential of which has yet to be assessed or exploited.[55]

RESPIRATORY PROTEINS

Membrane integrity is crucial for those enzymes concerned with energy conservation. Many of these such as cytochrome oxidase and ATP synthetase are intimately associated with the cytoplasmic membrane and rely on lipid for function. It follows that the thermotolerance of the respiratory chain and the ATPase will profoundly influence the efficiency of growth processes and indeed may limit the maximum temperature tolerated by an organism. Some aspects of respiratory proteins from thermophilic aerobic bacteria have been reviewed by Fee *et al.*[29] Kagawa and his co-workers[78] have analysed ATPase from a thermophile they designated PS3, which is likely to be a strain of *B. stearothermophilus*. They isolated a TF0–TF1 enzyme complex that was stable in urea at 70 °C. The purified TF1 was stable in organic solvents and more resistant to dissociating agents than mesophilic enzymes. Maximum activity occurred at 75 °C but could also be detected at 90 °C. For a comparison of thermophilic ATPase in PS3 with that from mitochondria or *E. coli* the reader is directed to Senior and Wise.[74] ATPase from the facultative thermophile *B. coagulans* has also been studied and, although not as thermostable as the PS3 enzyme, displays some activity even at 75 °C. Interestingly, the thermostability of the F1 portion is similar irrespective of growth temperature whereas the complete F0–F1 complex is more thermotolerant from cells grown at high temperatures. This implicates the membrane composition as being important in increased thermotolerance.[11,27] Properties of the respiratory chains of thermophiles have not been extensively studied. Two cytochrome *c* proteins and a cytochrome c_1aa_3 complex have been purified from *Thermus thermophilus*. The latter complex, which has a stoichiometry of c_1:aa_3 of 1:1, appears to be unique among bacterial cytochromes aa_3 oxidases.[89] The thermotolerance of the respiratory chain of *B. coagulans* reflected that of the primary dehydrogenases. NADH dehydrogenase exhibited intrinsic thermostability whereas succinate dehydrogenase activity could be stabilized by 10% NaCl only when bound to the membrane. Temperature for growth had a marked effect on respiratory chain composition. Cytochrome oxidase *d*, which was the major oxidase at 55 °C, could not be detected at 37 °C. This was shown to be due to the reduced oxygen tension as a result of the higher temperature which in turn induced the oxidase.[9,10] The effects of temperature on growth yields and the proportion of energy required for maintenance functions (such as upkeep of ionic gradients) have also been determined for some thermophilic bacteria grown in continuous culture. *B. caldotenax* yielded Y_{max} glucose values (gcell. mol glucose $^{-1}$) rising from 79 to 89 when grown between 60 and 70 °C.[52] In contrast, *Thermus thermophilus* values fell from 64 to 55 in a temperature range for growth between 60 and 78.5 °C.[62] Maintenance energy requirement of both these bacteria increased with rising temperature. Yield values for aerobic heterotrophic bacteria are known to be related to respiratory chain composition[46] and in the absence of detailed analysis of the cytochrome composition of thermophiles comparisons of yield data with other thermophilic or mesophilic species must be treated cautiously. Such analysis will be of prime importance if thermophilic species are to be used in industrial-scale fermentations. Yield of thermophilic anaerobes was investigated by growing *Thermoanaerobium ethanolicus* (69 °C) in continuous culture under glucose limitation. There was no significant difference in $Y_{glucose}$ or maintenance energy requirement when compared with a number of mesophilic anaerobes.[54]

Aerobic bacteria have evolved a number of enzymic systems to cope with toxicity of oxygen or its more reactive free radicals. Thermophiles in general have evolved in habitats of lower oxygen tension due to the decrease in gas solubility with rising temperature. The question of whether thermophiles require elaborate systems to protect against toxicity has been answered by a recent study. *T. aquaticus*, *T. thermophilus* and *Thermomicrobium roseum* all contained catalase, peroxidase and superoxide dismutase activities and could also tolerate increased oxygen tensions[60] implying that, for these bacteria at least, protective mechanisms against oxygen toxicity are still required.

tRNA

The thermostability of tRNA from *T. thermophilus* has been shown to be due to replacement of a G–U pair with a G–C pair as well as thiolation at T-55 and position 2 (uracil or thymine). In general, *T. thermophilus* contains T- (ribothymidine) and s2T- (2-thioribothymidine) tRNA. Their relative content depends on growth temperature. When active polysomes were prepared from cells grown at 55, 65 or 77 °C and the tRNA separated by HPLC it was found that there was more s2T-tRNA in polysomes from 77 °C-grown cells compared with 55 °C-grown cultures. Therefore the protein synthetic systems of the thermophile must have some selection mechanism to utilize either T- or s2T-containing tRNA.[85] The inclusion of s2T in tRNA serves to raise the melting temperature. The restricted conformation of the thiolated nucleotide as well as the strengthened stacking interaction that it allows between neighbouring bases also contribute to increased thermostability. This shows that relatively small changes in primary structure are sufficient to bring about increased thermostability.

POLYAMINES

Polyamines are important in regulation of nucleic acid synthesis, protein synthesis and cell division. They are abundant in actively growing cells and in bacteria their amounts decline in the stationary phase of growth. In mesophilic organisms only putrescine (diamine) and/or spermidine (triamine) are usually found. Thermophiles contain polyamines of much greater chain length. *B. stearothermophilus* synthesizes spermine as well as spermidine and putrescine. Their importance in mediating thermostability was illustrated in *T. thermophilus*. *In vivo* protein synthesis could not be demonstrated in cell-free extracts at 60 °C or higher. However, addition of certain polyamines restored the activity at 65 °C. Spermine was the most active of these. Further analysis of polyamines in *T. thermophilus* revealed twelve different types. Eight were new naturally occurring polyamines and two of these, homospermidine and norspermidine, were reported as being present in only a few organisms. It was found that the two major new components were tetramines called thermine and thermospermine, both of which could restore protein synthesis in the *in vivo* system described above for *T. thermophilus*. It was shown that thermospermine was most active in promoting thermotolerance and that it stabilized a ternary complex between the ribosomes, the messenger and aminoacyl-tRNA at high temperature. Thermine has also been detected in the caldoactive *S. acidocaldarius*. In addition, this thermophile contains small basic and abundant DNA-binding proteins that may have a role in maintaining the integrity of the DNA at high temperature.[36]

PROTEINS

Perhaps one of the most perplexing features of life at high temperatures is the way in which proteins, that to all intents are the same in amino acid composition as their mesophilic counterparts, are protected from denaturation. It is hardly surprising that this aspect has received most input from researchers, an area that has also been strongly supported by industry. This has stemmed from the need for stable long-lived enzyme molecules in a variety of industrial processes.

Thermotolerance of enzymes is a difficult term to define and is often referred to as the temperature optimum for activity. Describing an enzyme's thermostability in this way is misleading because two conflicting results of the effects of temperature are being observed. One is the effect on the rate of catalysis, the other on protein stability. Therefore highest rates of activity can be measured at a temperature that also causes a high rate of denaturation.[34] Probably the best way of describing an enzyme's thermotolerance is to select a temperature and quote the half-life (in unit time) of the protein at this value.

Numerous studies have been undertaken to identify universal mechanisms for promoting thermostability of proteins. These have ranged from comparison of sequence data for evolutionary well-conserved proteins in thermophiles and mesophiles to simple kinetic analysis of enzymes at different temperatures. This type of work has failed to identify any peculiar amino acid residues or protein conformation that can be correlated with enhanced thermostability despite comparisons between amino acid sequences and tertiary structures of well-characterized mesophilic proteins with their thermophilic counterparts. Another approach has been to compare the same protein that exhibits different thermostability in wild-type and mutant cells. This strategy has revealed that only one or a few amino acid changes are required for changes in the degree of heat tolerance to occur.

Factors affecting thermostability of enzymes in general

The tertiary structure is stabilized by forces such as hydrogen bonding, hydrophobic bonding, ionic interactions, metal binding and disulphide bridges. These either together or alone contribute to overall protein stability. No universal arrangement of any of these binding forces that endows thermotolerance on a protein from a thermophile can be recognized. Thermodynamic calculations suggest that the increase in free energy required to stabilize a thermophilic protein as compared to its mesophilic counterpart is in the order of only 5–7 kcal mol^{-1} which can be accomplished by a few hydrophobic bonds. This means that a relatively small alteration in enzyme primary structure, such as a single amino acid substitution, can lead to much more complex interactions at the tertiary level. Thermostability is also highly dependent on environment as shown by the fact that the majority of a protein's stabilizing forces are derived from interactions with the suspending medium. This is nicely illustrated by porcine lipase which is stable and active at 100 °C in an organic solvent. It follows that the observed thermostability of enzymes in aqueous laboratory solutions will probably be very different to those seen within the cell.

Mechanisms for promoting protein stability at high temperatures

Although there is no universal model for promotion of thermostability of proteins at high temperatures a number of different stabilizing mechanisms can be identified. It should be stressed that within a single thermophilic species different proteins will be protected by more than one mechanism. The following conclusions regarding the thermotolerance of proteins from thermophiles have emerged:

1. A single amino acid substitution is sufficient to alter the thermotolerance of a protein, e.g. the thermostability of tryptophan synthetase from *E. coli* was increased when a glutamate residue was replaced by methionine. This resulted in more hydrophobic bonding but raised the energy of stabilization of the enzyme molecule by only 3.14kJ mol^{-1}.[92]
2. The growth temperature can to some extent determine the thermotolerance of some proteins. Lauers and Heinen[56] found that in *Bacillus flavothermus* the heat stability of alanine dehydrogenase and some other soluble enzymes increased with growth temperature. Although the bacterium could grow in the range 34–75 °C, thermostability of the enzymes only changed up to 55 °C.
3. It is possible for a thermophile to synthesize a heat labile protein irrespective of growth temperature that has similar thermostability to a mesophilic one. The protein is stabilized at higher growth temperature by increasing the internal ionic bonding possibly by altering the internal ionic strength of the cell. Evidence for this comes from the studies of McLinden *et al.*[63] who studied the thermostability of glyceraldehyde-3-phosphate dehydrogenase from *B. coagulans*. The enzyme in crude cell extracts could be completely inactivated at the growth temperature of 55°C. Protection was afforded by increasing the ionic strength with neutral salts. Similarly, succinate dehydrogenase, a membrane-bound enzyme, from the same organism grown at 55 °C was rapidly inactivated at 60 °C but could be stabilized in the presence of 10% NaCl.[10]

4. Synthesis of intrinsically thermotolerant proteins, e.g. glyceraldehyde-3-phosphate dehydrogenase from *Thermus aquaticus* which is stable at 90 °C for several hours.[40] It is interesting in this respect that extremely thermostable enzymes can be isolated from species that are not particularly thermotolerant, e.g. α-amylase from *Bacillus licheniformis* (maximum temperature for growth approximately 50 °C) is active at 90–95 °C and is used extensively in the starch processing industry. Despite this, it is still most likely that thermophiles will be the main source of thermostable enzymes.
5. Some enzymes can be stabilized by substrates or effector molecules. Wedler and Hoffman[86] showed that glutamate synthetase from *B. stearothermophilus* could be stabilized in the presence of ammonia, glutamate and ATP at 65 °C — a temperature that normally inactivated it in the absence of these molecules.
6. Very often metals are able to stabilize proteins at high temperatures and indeed may be essential for activity. The classical example cited here is that of thermolysin, each protein molecule of which is stabilized by four Ca^{2+} ions.
7. In many fungi and yeasts polyhydric alcohols can enhance the thermostability of proteins, e.g. glycerol and sucrose are often used to stabilize enzymes during storage.
8. Chemical modification that involves alteration of the protein surface by chemical reagents or by covalent linkage to polymeric materials, as may be used during the immobilization of enzymes, may also improve thermostability.
9. Low water activity. There is much interest in enzymes prepared in mixtures of organic solvents and water. Initially work on activity in water and miscible solvents suggested that few enzymes functioned when the concentration of the solvent exceeded 50% (w/v). This concentration was insufficient to solubilize water-insoluble substrates or products which this type of system would be eminently suited. It was assumed that the same applied to water-immiscible solvent mixtures but recent experiments have disproved this, at least for some enzymes. The best example is afforded by pancreatic porcine lipase. At 100 °C in phosphate buffer this enzyme is almost instantaneously denatured. If the enzyme is added as a powder to 2 M *n*-heptanol in tributyrin that contains 0.8% water, 50% activity is retained after 50 min and almost 100% after 60 min if the water content is reduced further to 0.015%.[34] Similar work has been reported for two membrane-bound enzymes from beef heart submitochondrial particles. Cytochrome oxidase and ATP synthetase could be transferred into highly apolar solvents as protein–lipid complexes. At a water concentration of 3 μl ml^{-1} toluene the half-life of ATPase at 90 °C was 5 h.[8] This type of work has obvious commercial potential for those water-insoluble substrates or products such as lipids and also stresses the importance of water activity in promoting thermal denaturation of proteins.
10. Thermophily imposes a necessity for different levels of control on activity of some enzymes, e.g. L-asparaginase from *B. coagulans*, *B. stearothermophilus* and *Thermus aquaticus* differs from that of mesophilic species by having high K_m values and high substrate specificity.[23]

Some rules for increasing the thermostability of proteins

Arising from numerous studies in which thermophilic proteins have been compared with their mesophilic or thermolabile counterparts, it appears that to improve the inherent thermostability of proteins all that is required is the replacement of one or more key amino acid residues. The net effect aimed for should be to increase the various bonding forces within the protein molecule which in turn increase the energy of stabilization of the molecule. Of course the crucial decision is which amino acid residues to replace. In general, single replacements should: (i) be in those on the external parts of the molecule; (ii) preferably be in beta-turn rather than helix or sheet; and (iii) conserve the secondary structure of the protein, especially when the residue is in an alpha helix or beta sheet. The following replacements are preferred in this order to increase heat tolerance: Asp to Glu, Lys to Gln, Val to Thr, Ser to Asn, Ile to Thr, Asn to Asp.[71]

Argos *et al.*[6] compared the amino sequences of glyceraldehyde-3-phosphate dehydrogenase, lactate dehydrogenase and ferredoxin from thermophiles and mesophiles. They deduced that for these proteins that the following replacements were most common in the thermophiles: Gly to Ala; Ser to Ala; Ser to Thr; Lys to Arg; and Asp to Glu. The net effect of such changes is to: (i) increase the volume of amino acids in the thermophilic proteins which could result in more compact packing; (ii) increase the internal and decrease the external hydrophobicity; and (iii) stabilize alpha helical structures. A higher content of arginine and a lower one of lysine has also been reported for thermostable proteins. This observation has been explained by the fact that nearly all the lysine and arginine residues are on the surface of the protein molecule and in contact with water. Lysine has a longer hydrocarbon chain for which contact with water is thermodynamically unfavourable. The area of contact for arginine is less because the chain is one CH_2 shorter.

Increased hydrophobicity has also been proposed to correlate with thermostability but is thought to be more important for amino acid residues that are located internally. A recent report on RNAse from *Bacillus amyloliquefaciens* showed that truncated hydrophobic side chains produced by site-directed mutagenesis resulted in an increase in the susceptibility to urea denaturation[47] although the effect on thermotolerance was not investigated.

Replacement of single or few amino acid residues

The amino acid sequence of ferredoxin in *Clostridium tartarivorum* and *C. thermosaccharolyticum* are essentially the same except that for the former residues 31 and 44 are glutamine but in the latter are glutamic acid. Perutz and Raidt[69] considered that this difference allowed the formation of two extra salt bridges between lys 29 and glu 31 and between his 2 and glu 44 and that this was sufficient to account for the difference in thermostability between the two clostridial enzymes. Ferredoxin from *C. tartarivorum* lost 50% of its activity at 70 °C after 2 h whereas that from *C. thermosaccharolyticum* lost only 10% of its activity in the same time. Tryptophan synthetase from a mutant strain of *E. coli* was shown to be more thermostable than that of the wild-type. This was shown[91] to be due to the replacement of a hydrophilic residue (glu 49) with a hydrophobic one (methionine). The *npr* T gene that encodes a thermostable neutral protease (Npr T) from *B.*

stearothermophilus CU21 has been cloned and sequenced. This has been used to show that the amino acid sequence has 85% homology with that of thermolysin although the latter is more thermostable.[33,83] Imanaka *et al.*[44] were able to enhance the thermostability of Npr T by site-directed mutagenesis although not to the level of thermolysin.[51] Suzuki *et al.*[82] recently reported a strong correlation between increase in proline content and a rise in thermostability of oligo-1,6-glucosidase in thermophilic species of bacilli of different thermotolerance.

Commercial aspects

Many advantages for the use of thermophiles in biotechnological processes have been proposed.[75,79,88] Some of these are listed here and, where appropriate, their true benefit discussed.

1. *Faster reaction times.* In terms of growth rate there are many mesophilic species that grow more rapidly than many thermophiles. In addition many of the extreme thermophiles require specialized growth factors such as polypeptides in order to grow at high temperatures. However, there are distinct advantages for using thermophilic species that produce secondary metabolites as long as they grow faster than their mesophilic counterparts. In processes where the end-product is a secondary metabolite it may be a matter of days or considerably longer before the culture reaches the producing phase. This growth-related delay may be shortened in thermophilic processes.
2. *Reduced risks of contamination.* Due to the large numbers of spores of both mesophilic and thermophilic species (notably *Bacillus*) found in industrial growth media this may not always be an advantage in practice.
3. *Reduced cooling costs of operating large fermentations.* It has been calculated that the optimal temperatures for operating bioreactors that minimize energy input as cooling or heating is around 45 to 50 °C which would appear to favour the use of the thermotolerant or facultatively thermophilic species. Energy costs of heat input to maintain extreme thermophiles may be as costly as that required to cool the mesophilic ones.
4. *Viscosity.* The viscosity of growth media decreases with increasing temperature which in turn reduces the energy requirement for agitation and allows better mixing, especially in large fermentation vats.
5. *Ionization and solubility.* Most inorganic and organic molecules show increased ionization and higher solubility at high temperatures so that there is potential for greater production of biomass and/or product. However, the solubility of oxygen decreases with a rise in temperature which restricts the temperature for growth of the aerobic thermophiles. Conversely, this is advantageous to anaerobic thermophilic processes.
6. *Pathogens.* Thermophilic treatment of sewage wastes will kill off pathogenic bacteria and viruses.
7. *Distillation of volatile end-products.* Operations at high temperature have been proposed as a means of microbial production of volatile chemicals that can be

distilled off at the temperature of growth and thus allow cheap, rapid purification of product whilst at the same time removing any inhibition of microbial growth due to accumulation of the end-product. A preliminary study is under way to assess the feasibility of ethanol production by a *B. stearothermophilus* isolate. Other advantages of thermophilic enzymes are increased resistance to chemicals, a different substrate spectrum than comparable mesophilic proteins, different stereospecificity, longer life and less chance of contamination if operated at sufficiently high temperature.

There are also some disadvantages in the use of thermophilic processes. Existing facilities for large-scale fermentations are not suited to the cultivation of the extremely thermophilic species. High temperatures will increase thermolability of medium constituents and there will be increased loss of water due to evaporation. Developments in the commercial exploitation of caldoactive and barothermophiles will require advances in handling and cultivation methods.[24] Finally, the genetic methods applicable to thermophiles are at present limited making strain improvement difficult.

INDUSTRIAL ENZYMES

Table 1.2 summarizes some of the major enzyme activities important to industry and details those thermophilic examples that have been characterized to date. At the outset it has to be said that industrial conversions are mostly mediated by enzymes from mesophiles, and that the bulk of investigations have centred on *Bacillus* and *Thermus*. Proteases are probably the most widely used enzymes commercially. They fall into one of three categories: acid (pH 2–4), neutral (pH 7–8 and inhibited by metal chalators) and alkaline (pH 9–11) proteases. Their industrial use will therefore vary and depend on pH of the bioconversion and also means that as a group they have wide application. The calcium-stabilized proteases are unsuitable for use in biological washing powders as these contain calcium binding agents. Of the thermophilic proteases the most studied is thermolysin from *Bacillus thermoproteolyticus*. This is a neutral protease which is stabilized by four Ca^{2+} ions and produced by growth of the bacterium at around 55 °C which in itself is not a particularly thermophilic temperature. Therefore we should note here that thermostable proteins need not necessarily come from thermophilic species. This is also illustrated by the example of glucose isomerases produced by thermophilic and thermotolerant Bacilli which exhibit reasonable thermostability but not as much as the enzyme from a mesophilic *Actinoplanes* which retains activity at 90 °C for 20 min. However, the fact remains that there is a much greater chance of finding thermostable proteins from thermophilic organisms. Caldolysin is much more thermostable than thermolysin and in addition is extremely stable towards denaturing agents. There is evidence that immobilization of this protease markedly improves its thermotolerance, for example when immobilized in CM-cellulose or Sepharose the half-life at 85 °C increased from 360 to 1060 min and at 95 °C from 28 to 128 min. Other enzymes immobilized in similar ways also showed elevated thermotolerance and this may be a promising method for improving thermostability of proteins as long as activity is unaffected or

Table 1.2 Examples and properties of some commercially important enzymes from thermophiles (taken principally from refs 17, 75 and 88)

Enzyme	*Source*	*Thermotolerance*	*Applications*
Proteases			
Thermolysin (neutral protease)	*B. thermoproteolyticus*	$t_{\frac{1}{2}}$ 1 h at 80°C, stabilized by four Ca^{2+} ions	Proteases used in leather industry, brewing, baking, cheese processing
Aqualysin I (alkaline protease)	*T. aquaticus*		Detergents — biological washing powders
Aqualysin II (neutral protease)	*T. aquaticus*		
Caldolysin (stable pH 4–12 at 75 °C)	*T. aquaticus*	$t_{\frac{1}{2}}$ 30 h at 80°C stabilized by six Ca^{2+} ions	
Cellulases			
Cellulases Hemicellulases Cellobiases	*C. thermocellum* Thermophilic actinomycetes e.g. *Thermomonospora* sp. Thermophilic *Bacillus* sp.	Stable at 70 °C 15 min Optimum temp. 65–70 °C Stable 70 °C for 15 min	Release of sugars from cellulose (β-1, 4-D-glucose) and hemi-cellulose (β-1, 4 xylose) found in wastes from agriculture, paper making and anaerobic digesters
Amylases			
α-amylases β-amylase	Thermophilic *Bacillus* actinomycetes *B. licheniformis*	Usually stabilized by Ca^{2+} ions Active at 90–95 °C	α-1, 4 glucosidic linkages, in amylose, amylopectin and glycogen. Used in brewing, the end product of β-amylase is maltose
Others			
Glucose isomerase	*B. coagulans*		Glucose to fructose as sweetener in food industry
	B. stearothermophilus	Stable at 80 °C for 30 min	
Alcohol dehydrogenase	*Thermoanaerobium ethanolicus*	Little loss of activity at 70 °C over 2 days	Redox reactions using organic solvents as substrates
β-Galactosidase	*T. aquaticus*	$t_{\frac{1}{2}}$ 8 min at 90°C (10% loss after 36 days at 65°C)	Hydrolysis of lactose (e.g. milk whey) syrups to glucose and galactose
	B. stearothermophilus	$t_{\frac{1}{2}}$ 20 days at 50°C	
	Caldariella acidophila	$t_{\frac{1}{2}}$ 30 days at 70°C in an immobilized form	

Table 1.2 cont.

Enzyme	*Source*	*Thermotolerance*	*Applications*
Restriction endonucleases			
*Bst*II	*B. stearothermophilus*	Gene cloning	
*Bcl*I	*B. caldolyticus*		
*TEH*III	*T. thermophilus*		
*Taq*I	*T. aquaticus*		
Hydrogenases	*Methanobacterium thermoautotrophicum*	No loss of activity over 2 weeks at 25 °C under H_2	Production of H_2 Regeneration of NADH or NADPH
Glycerokinase	*B. stearothermophilus*	$t_{\frac{1}{2}}$ 310 min at 60°C	Estimation of serum triglycerides
Lipases			Detergents, food processing

as in the case of caldolysin improved. Immobilizing the thermophile PS3 in a glutaraldehyde bovine serum albumin mixture resulted in the cytoplasmic membrane becoming permeable to NADH. Studies of the cycling of NADH to NAD^+ by the immobilized cells showed them to be some ten times better than *E. coli* and much more stable.[28] An interesting observation regarding the binding of *B. stearothermophilus* to starch and related glucans was made by Ferenci and Lee.[30] At 55 °C there was little association but at 25 °C the bacterium bound strongly to the polymer. This could be a novel survival mechanism whereby attachment to a solid substrate can occur under non-optical growth conditions for possible utilization upon a shift to more favourable ones.

Extracellular cellulase from *Clostridium thermocellum* has no associated proteolytic activity and functions unaffected by calcium, magnesium, or manganese at concentrations that inhibit the cellulase from the fungus *Trichoderma*. It is not subject to end-product (glucose) inhibition or affected by low concentrations of cellobiose (unlike the fungal enzyme) and it has a much higher activity. Recently a thermostable xylanase from the thermophilic fungus *Thermoascus aurantiacus* has been reported which has a half-life of 1.5 h at 70 °C.[90]

A number of endonucleases has also been found in thermophiles, e.g. *Taq*1 from *T. aquaticus* recognizes the four-base sequence 5′-TCGA-3′/3′-AGCT-5′ and cleaves between T and C. A restriction endonuclease *Tth*111 I has been isolated from *T. thermophilus* that recognizes the sequence GACN/NNGTC and cleaves at the hyphenated part. This enzyme like *Taq*I is heat stable.

Allais *et al.*[3] isolated four thermophilic *Bacillus* species that had thermostable beta fructosidase (inulinase) activity and proposed that they had potential for the production of high fructose syrups from inulin which is a linear polymer of fructose of plant origin.

A number of thermophilic hydrocarbon-utilizing bacteria have also been isolated. Enrichments in *n*-alkane containing media at 60 °C yielded Gram-negative bacteria. Enrichments with methane and propane were unsuccessful despite the wide occurrence of mesophilic species that are able to use these gases. The Gram-negative isolates were non-pigmented, non-spore forming immobile rods that were obligately thermophilic (optimum temperature for growth 55–65 °C). No Gram-positive thermophilic hydrocarbon utilizers could be isolated despite the numerous mesophilic species that are known to do so.[68]

Alcohol dehydrogenases from thermophiles have a broad substrate spectrum. The *Thermoanaerobium brockii* and *Clostridium thermohydrosulfuricum* enzymes show maximal activity with secondary alcohols and least activity with primary alcohols. They also show high tolerance to solvents. *Thermoanaerobium ethanolicus* has primary and secondary alcohol dehydrogenases, both of which have been purified to homogeneity and are highly thermostable.[18]

A number of thermophilic bacteria capable of growth on CO as a sole carbon source have been identified. These include bacilli[50] and a thermophilic streptomycete.[13] The CO oxidase of the latter appeared to be a molybdenum hydroxylase similar to the enzymes from the mesophilic carboxydotrophs. Recently a thermostable peroxidase, enzymes widely used clinically for colorimetric measurements of biological compounds, has been purified from *B. stearothermophilus*.[58]

WASTE TREATMENT AND METHANE PRODUCTION

Anaerobic digesters of domestic and agricultural sewage have been proposed as a cheap means of obtaining energy in the form of methane gas. The overall reaction results in the anaerobic breakdown of complex organic solids to CO, H_2 and CH_4 to leave a much reduced solids residue. The process is mediated by a consortium of ill-defined anaerobic bacteria that also comprises many thermophilic species especially amongst the methanogens. To date most anaerobic digesters have been run at mesophilic temperatures but there is much interest and potential in adopting a thermophilic system. Advantages that have been proposed are: (i) faster rates of conversion of organic matter to CH_4 which in turn means reduced retention times of the solids; (ii) increased throughput and improved efficiency; (iii) lower biomass produced due to the lower growth efficiency of the thermophiles; (iv) growth of pathogenic bacteria and viruses inhibited; (v) CH_4 can be used as a fuel; (vi) no requirements for aeration which is expensive; and (vii) lower viscosity at the higher temperatures which reduces the energy required for mixing. There are also some disadvantages to using such a system. For example, the few processes that have been set up have shown some tendency to instability, particularly when the nature or the size of the loading is changed; for some systems the residence time of the solids has in fact been longer; finally, there is some requirement for an input of energy as heat in order to ensure a constant operating temperature.

Methanogenic bacteria have a range of substrates that is limited to H_2, CO_2, acetate and formate. Therefore, in anaerobic digesters, a community of anaerobes that exhibit a range of metabolic capabilities is necessary to degrade complex organic matter such as polymeric materials like cellulose, polysaccharides and proteins. First the fermentative species that possess the requisite enzymes (invariably extracellular) act to attack macromolecules and convert them to simpler monomers and oligomers such as acetate, alcohols, fatty acids, H_2 and CO_2, e.g. *C. thermocellum* converts cellulose to hexoses and ethanol, lactate, acetate,

H_2 and CO_2; *C. thermohydrosulfuricum* and *Thermoanaerobium brockii* convert amino acids and sugars to acetate, ethanol, lactate, H_2, CO_2 and long-chain fatty acids (longer than acetate). Hydrogen-evolving acetogenic bacteria oxidize propionate, butyrate and other fatty acids to acetate, H_2 and CO_2. These bacteria require low hydrogen tensions although they often live in habitats that have high and continuous rates of hydrogen evolution. This paradox may be resolved by the discovery of thermophilic obligately autotrophic H_2-oxidizing bacteria such as *Hydrogenobacter autotrophicus* and *Calderobacterium hydrogenophilum*.[15] The end result of these activities is the fermentation of complex macromolecules to H_2, CO_2, acetate and formate that are converted to methane by the methanogens. These include *Methanobacterium thermoautotrophicum* (H_2 and CO_2 to CH_4); *Methanococcus thermolithotrophicus* (H_2, CO_2 and formate to CH_4 and some CO_2) and *Methanothrix* sp. (acetate to CH_4 and CO_2). Methanogenesis in defined co-cultures has been studied on fermentations of sugar beet pulp. *C. thermocellum* alone produced acetate, succinate, methanol, ethanol, H_2 and CO_2. A co-culture with a species of *Methanobacterium* resulted in only trace amounts of ethanol and succinate. Acetate concentration was about three times that produced by *C. thermocellum* alone. Association of this co-culture with *Methanosarcina* resulted in 75% of total carbohydrate in the pulp being converted to methane.[67]

Usually anaerobic reactors have been operated at mesophilic temperatures (35–40 °C) despite the knowledge that they can also be operated at 45–65 °C. Propionate and butyrate account for around 20% of the CH_4 produced in these environments.[1] A recent study indicated that operating temperature plays an important part in determining the microbial flora. At 37 °C sewage digestion was largely mediated by asporogenous bacteria whereas the same reactor at 55 °C was dominated largely by spore formers.[20] Major conclusions from the different reactors that have been run at high temperature are as follows:

1. Temperatures greater than 50 °C resulted in rapid rise in volatile acids.
2. Retention time was reduced from 14 to 7 days in one process due to more rapid reactions at high temperature. Higher rates of biodegradation allowed a two-fold increase in loading at thermophilic conditions and improved throughput.
3. Better and more rapid gas production was achieved.
4. Temperatures above 60°C promoted adverse effects. The performance of thermophilic methane-producing digesters has also been monitored by analysing total lipids. Chemical measurements of lipids appear to offer a quantitative way to correlate shifts in microbial biomass, community structure and nutritional status. In addition high levels of polyhydroxybutyrate were indicative of metabolic stress.[38]

COMPOSTING

This is an aerobic process that degrades moist, solid organic wastes in a process that starts off at mesophilic temperatures but due to self-heating that results from microbial activity may reach thermophilic conditions as temperatures rise up to

Table 1.3 Numbers of thermophilic bacteria recovered from compost (W. Amner, unpublished results)

Isolation medium	*Approximate numbers of thermophilic bacteria g compost^{-1}*				
	Thermomonospora	*Thermoactinomyces*	*Thermomonospora chromogena*	*Streptomyces*	*Bacillus*
Half-strength tryptone soya agar + 0.2% casein ($\frac{1}{2}$ TSA)	10^7	10^5	10^6	10^5	10^{10}
$\frac{1}{2}$ TSA + 5% NaCl	10^6	—	—	10^5	10^7
Half-strength nutrient agar	10^7	10^4	10^7	10^7	10^8
Compost infusion agar	10^8	10^{10}	—	10^7	10^8

80 °C. This is reflected in the microbial composition of compost that is mediated by both mesophilic and thermophilic organisms that are chiefly Gram-positive and include *Bacillus*, *Streptomyces*, *Thermomonospora*, *Saccharomonospora* and *Thermoactinomyces*. Table 1.3 shows the distribution and approximate numbers of these thermophilic species isolated from phase two mushroom compost using a number of different solid media. It is apparent that thermophilic Bacilli dominate and more selective methods are required to isolate the actinomycetes. Also, different media allow the proliferation of some species in preference to others. Fungi such as *Aspergillus fumigatus* which is an opportunistic pathogen and thermotolerant may also be encountered. The microbial composition of compost during the thermophilic phase has been studied by Strom.[80] The end-product is a stable humus-like product that is suited for use as a low C/N fertilizer. Preparation of compost is a well-standardized process in the mushroom industry. Raw materials typically include horse or chicken manure that are stacked in piles and turned occasionally. Due to microbial activity the temperature rises to about 70 °C and this phase is therefore characterized by a succession of mesophilic species to thermophilic ones. Phase two is a pasteurization stage in which the compost is rapidly heated to 60 °C, then 50°C for 7 days during which the actinomycetes and Bacilli become increasingly dominant and also most of the ammonia is driven off at this time. The effect of temperature on the composting of sewage sludge has been investigated at 50, 60 and 70 °C. Total CO_2 evolution and final conversion of volatile matter was maximal at 60 °C. The dominant thermophilic bacterium at this temperature was isolated and shown to have a CO_2 value at 70 °C that was four times that at 60 °C implying that anabolism and catabolism become uncoupled at the higher temperature.[66]

LIGNOCELLULOSE DEGRADATION

There is increasing interest in using plant biomass as a potential energy source, for example by completely degrading straw using microorganisms and releasing the constituent sugar monomers via a biological saccharification process. The released carbon is then used in secondary fermentations for production of high volume low cost products such as ethanol or low volume high cost products such as pharmaceuticals. Such a programme is reliant on the cheap and efficient degradation of cellulose and hemicellulose as well as a means for removing lignin from lignocellulosic materials in order to allow a more rapid degradation of the cellulosic polymers. There are many examples of microorganisms that will degrade either cellulose or hemicellulose and many enzymes (cellulases, xylanases and xylosidases) have been isolated. Lignin is by far the most recalcitrant of the plant polymers and most of the early work has concentrated on fungal systems, notably *Phanerochaete chrysosporium*. Recently attention has focused on actinomycetes as potential agents for degradation of lignocellulosics. In particular thermophilic actinomycetes have received especial interest. Their role in the environment in the breakdown of plant matter is recognized as well as their importance in composting (see ref. 61 for review).

ETHANOL PRODUCTION

Production of fuels by thermophiles relies on a cheap and readily available supply of carbohydrate. One prime candidate is that derived from plant biomass such as wood residues (forestry, pulping for paper making), straw and other cellulosic wastes. These have the advantage in that they are renewable energy sources unlike the fossil fuels. The major interest has been in the production of ethanol from carbohydrates. This is a traditional fermentation in which commercial strains of yeast are able to produce high yields of ethanol but the process suffers from an applied point of view because a lot of biomass is produced at the same time. Much interest has been expressed in using thermophiles for ethanol production because it may be possible to distill off the ethanol continuously at the growth temperature of the producing organism. Such an approach would allow better conversion of substrate to product, decrease the energy input into the fermentation (low cooling costs) and make downstream processing steps more economic since there would be no requirement for removing the cells from the fermentation broth in order to recover the ethanol.

One candidate for such a process is *C. thermocellum* which is strictly anaerobic and can ferment cellulose to produce a mixture of ethanol, acetate, lactate, H_2 and CO_2. The degradation of cellulose requires an exocellulase (β-1,4-D-glucan cellobiohydrolase) and endocellulase (β-1,4-D-glucan hydrolase) which act co-operatively to depolymerize cellulose to cellobiose and oligosaccharides and a β-glucosidase which hydrolyses these to form glucose.[59] However, the yields of ethanol are currently low due to incomplete substrate utilization, the formation of additional end-products and a poor tolerance to ethanol. Attempts to circumvent these problems have been made by co-culturing *C. thermocellum* with other thermophilic anaerobes that, although incapable of degrading cellulose, are able to utilize monosaccharides left by the activity of *C. thermocellum*. One such thermophile is *Thermoanaerobium ethanolicus* and it has a similar growth temperature range (37–78 °C) to that of *C. thermocellum* (45–65 °C). An advantage of such co-culture experiments is that removal of saccharide products of cellulose breakdown by the partner bacterium actually stimulates the cellulolytic species to degrade the cellulose faster.[81] Some thermophilic anaerobes are also able to utilize pentose sugars (derived from hemicelluloses) such as xylose. For example, a strain of *T. ethanolicus* was able to produce 1.4 mol ethanol mol xylose^{-1} at low substrate concentrations. However, it has also been noted that other strains of *T. ethanolicus* gave high ethanol yields on individual monosaccharides (e.g. glucose) but much reduced ones on carbohydrate mixtures (e.g. glucose/starch/xylose). Thus it would appear that the co-culture method outlined above is not universally suitable. Such experiments throw up interesting questions regarding the control and regulation of metabolic pathways as well as ecological interactions. Sissons *et al.*[77] isolated eight anaerobic cellulolytic cultures from thermal pools in New Zealand. The cellulase activity of one isolate which they named as *Caldocellum saccharolyticum* was more thermostable than *C. thermocellum*. Production of ethanol from cellulosic biomass poses operational difficulties. Some pre-treatment of the substrate is required before use; the insolubility of the biomass will influence the substrate loading since degradation rates will tend to be slower than for soluble substrates. To date such fermentations do not appear to yield ethanol yields any greater than 4 g l^{-1}.

Another approach has been to develop a strain of *B. stearothermophilus* to produce ethanol. This bacterium normally produces only small amounts of ethanol but a mutant that lacks lactate dehydrogenase has been isolated which shows a dramatic improvement in yield to 1 mol ethanol mol glucose^{-1} fermented. Work is now in progress to improve the yields further in view of the known high tolerance of this organism to ethanol, together with the fact that cloning work is best developed in this thermophile compared with others.

Genetic aspects

MOLECULAR STUDIES

DNA cloning and site-directed mutagenesis have opened up new possibilities for studies of thermophiles and their products. These techniques make it feasible to design proteins that are more thermostable than even their thermophilic counterparts. Genetic studies in thermophiles have largely been confined to *B. stearothermophilus*, *Thermus thermophilus* and to some degree *Sulfolobus acidocaldarius*. Many studies have concentrated on developing cloning strategies by screening for suitable vectors, identifying thermostable markers and using these in gene transfer experiments from thermophile to thermophile, thermophile to mesophile or mesophile to thermophile. This work has proved valuable for understanding gene expression in host backgrounds of differing heat sensitivity and also for increasing our understanding of the mechanisms of thermotolerance.

PLASMID VECTORS

Table 1.4 lists some of the plasmid vectors that have been used in cloning experiments primarily within the *Bacillus* group. This genus is particularly favoured because genetic exchange mechanisms have been well worked out for

Table 1.4 Properties of some plasmid vectors isolated from thermophilic bacilli

Plasmid	*Source*	*Molecular weight (MDa)*	*Characteristics*	*Notes*	*Reference*
pUB110	*Staphylococcus aureus*	3.0	Kan^r	Expressed in *B. stearothermophilus* up to 55 °C. Unstable at higher temperatures	53
pAB118A	Thermophilic bacilli	4.9	$Strept^r$	No expression in *B. subtilis*	2
pAB124	Thermophilic bacilli	2.9	Tet^r	Successful transformation of *B. subtilis*. Potential as shuttle vector between thermophilic and mesophilic bacilli	14
pTHT9	Thermophilic bacilli	5.4	Tet^r	Extensive homology in the regions that include replication origin and tet^r	41
pTHT15	Thermophilic bacilli	3.2	Tet^r		
pTHT22	Thermophilic bacilli	5.9	Tet^r		
pTHN1	Thermophilic bacilli	3.4	Kan^r	Replicates in both *B. subtilis* and *B. stearothermophilus*. Potential as shuttle vector	
pTB19	Thermophilic bacilli	17.2	Kan^r, Tet^r	Expressed at 65 °C but some loss from a proportion of the host cells	43
pTB20	Thermophilic bacilli	2.8	Tet^r	Restriction analysis suggests similarity to pAB124	
pTB53	Thermophilic bacilli	11.2	Kan^r, Tet^r	Used in cloning of neutral protease from *B. stearothermophilus* into *B. subtilis*	51

mesophiles[19] and these appear to be applicable to the thermophilic species. In addition, *Bacillus* comprises many industrially important species and as a group they characteristically produce many extracellular proteins. This makes them good recombinant hosts for secretion of cloned proteins out of the cell. Many of the plasmids shown in Table 1.4 have reportedly been successfully used to transform *B. stearothermophilus* (e.g. pUB110, pTB19, pTHT9). pTHT15 and pTHN1 each express in both *B. subtilis* and *B. stearothermophilus* even in the absence of selective pressure, making them ideal shuttle vectors between thermophile and mesophile. pTB20 and pAB124 which both encode tetracycline resistance have been shown to be very similar by restriction analysis.[41] Different strains of *T. thermophilus*, which grows at temperatures above 75 °C, have been shown to contain plasmids that are cryptic. The lack of plasmid vectors that carry suitable markers has hampered the development of transformation systems in this thermophile. The high growth temperatures may be prohibitive to the acquisition of antibiotic resistance genes.

THERMOPHILIC BACTERIOPHAGES

Sharp *et al.*[76] isolated twenty-four thermophilic bacteriophages that were able to infect most of the major groups of *Bacillus* thermophiles. The phages varied in size and shape and most were stable at 50°C for 4–5h but at 70°C there was a dramatic reduction of between 10^2 to 10^7-fold in 2 h. The DNA profiles obtained after restriction analysis showed different banding patterns even in morphologically similar phages. A restriction–modification system could be demonstrated for *B. caldotenax*. Finally, transduction of *B. caldovelox* and *B. caldotenax* by a phage designated JS017 indicated that thermophilic phages may be useful for studies of the genetics of thermophilic bacilli.

TRANSFORMATION

Transformation has been used for promoting genetic exchange in thermophilic bacilli, particularly in *B. stearothermophilus*. The system developed for *B. subtilis* requires some modifications for the thermophile. The changes include using 1 μg ml^{-1} lysozyme for protoplast formation (rather than 2 mg ml^{-1} for *B. subtilis*) in order to enhance regeneration of protoplasts; because *B. stearothermophilus* dies quickly at room temperature but is stably maintained at 4 °C all procedures are done either at the latter temperature or 48 °C; finally sucrose is used (succinate used for *B. subtilis*) in the agar used for regeneration of protoplasts. Plasmid vectors could be prepared from *B. subtilis* or *B. stearothermophilus* and used to transform the same recipient. In this way transformation efficiencies of a vector which had undergone thermophilic or mesophilic conditioning before its reintroduction into the host bacterium could be compared. It was found that pTB19, pTB90 or pTHT15 prepared from *B. subtilis* resulted in transformation frequencies that were reduced by two- to three-fold when compared with those using the same plasmids prepared from *B. stearothermophilus*. Another plasmid (pUB110) yielded similar transformation frequencies irrespective of the way in which it was prepared. It was noted that pTB90 gave the highest numbers of transformants of the thermophile. Temperature had a pronounced effect on the expression of the plasmid in transformed cells. pUB110 was stably maintained at up to 55 °C but became unstable at temperatures above this, although, even after growth at 65 °C for twenty generations, it could be shown that 10% of the population still carried the plasmid, implying that the plasmid replication was not totally heat inactivated. pTB90 (tet^r and kan^r) was also stably maintained at 55 °C but became increasingly unstable at 60 and 65 °C. However, *B. stearothermophilus* could still grow in the presence of kanamycin or tetracycline even at 65 °C. It is somewhat surprising that, in view of the detailed published methodologies listed for transforming *B. stearothermophilus*, not more work has been published in this area.

An alternative strategy has been developed for *Thermus*. Genetic transformation of *T. thermophilus* auxotrophs to prototrophy was achieved at high frequency when growing cells were exposed to chromosomal DNA from a nutritionally independent strain. No chemical treatment was required to induce competence and the cells were receptive at all stages of growth. Highest frequencies were obtained at pH6–9 and 70 °C. Other strains, that included *T. flavus*, *T. caldophila* and *T. aquaticus* could be transformed to streptomycin resistance by DNA from their own spontaneous streptomycin resistant mutants. Finally, a cryptic plasmid was introduced into *T. thermophilus* K102 by adding 50 μl of culture to 5 μg dried plasmid DNA, incubating for 1 h and plating out. A screen of the resultant colonies for the acquisition of the plasmid produced 9 colonies per 1000 tested.[48]

The problem of the thermostability of the resistance markers of plasmid vectors has been assessed for kanamycin resistance as described above in pTB913, a derivative of pTB19. The gene product that promotes kanamycin resistance has been identified as an enzyme kanamycin nucleotidyl transferase which is similar to that encoded by pUB110 from a mesophile *Staphylococcus aureus*. The enzyme encoded by pTB913 was more thermostable than that from pUB110 as shown by the fact that approximately 65% of activity remained after heating at 55 °C for 5 min compared with only around 15% for that from pUB110. Some sequence analysis of DNA showed a base substitution at position 389 from cytosine (pUB110) to adenine (pTB913) which resulted in threonine at position 130 being replaced by a lysine for the kanamycin nucleotidyl transferase coded for by the pTB913 plasmid. This also furnishes another example of how a small change in amino acid sequence results in increased overall thermostability of a protein.

MESOPHILIC GENE INTO THERMOPHILIC HOST

Using plasmid pTB90, a derivative of pTB19 (see Table 1.4), Fujii *et al*[32] were able to clone the penicillinase genes from a mesophile (*B. licheniformis*) into a thermophile (*B. stearothermophilus*). The genes for this extracellular enzyme were expressed in the thermophile even at 60 °C. The amount of enzyme produced at 48 °C in the recombinant was the same as that produced at 37 °C by *B. licheniformis*. Above 48 °C the amount of enzyme produced decreased. Interestingly, the enzyme was predominantly extracellular in the clones indicating that there are common protein secretion systems amongst the bacilli.

THERMOPHILIC GENES INTO A MESOPHILIC HOST

An early study involved the cloning of the structural gene for 3-isopropylmalate dehydrogenase (3-IPM dehydrogenase) from *Thermus thermophilus* into the mesophile *E. coli*.[21] This gene corresponds to the *E. coli leu*B gene. The strategy used involved mixing plasmid pBR322 with 1 μg DNA from the thermophile and digesting the mixture with *Hind*111. DNA fragments from the thermophile and plasmid were ligated and used to transform an *E. coli leu*B mutant and selecting for transformants that were ampicillin resistant and *leu*$^+$. Analysis of transformants that now carried the recombinant plasmid pHB2 revealed that they possessed the thermophilic genes for isopropylmalate dehydratase (*leu*C product) and 3-IPM dehydrogenase. *E. coli* cells were now able to produce 3-IPM dehydrogenase and isopropylmalate dehydratase enzymes that had similar heat resistance to those synthesized by the thermophile. They could be easily purified from *E. coli* by heating cell extracts at 70 °C for 5 min, a time sufficient to denature most *E. coli* proteins which could then be easily removed by centrifugation leaving unaffected thermophilic enzymes in the supernatant. A gene library was prepared using DNA from *Caldocellum saccharolyticum* (grows at 80 °C) and subsequently used to clone the gene for beta-glucosidase into *E. coli* and *B. subtilis*. The enzyme was cell associated in both host bacteria and showed the same thermostability as seen in *C. saccharolyticum*. Hirata *et al.*[39] cloned the β-galactosidase gene from *B. stearothermophilus* into *B. subtilis*. Normally the enzyme is induced at low levels by lactose. However, it was produced constitutively in the mesophilic host and its production was enhanced fifty-fold. A further advantage was the ease by which it could be purified from *B. subtilis*. Heating cell extracts of the host at 70 °C for 15 min and removal of denatured proteins by centrifugation resulted in a purification to 80% homogeneity. These experiments illustrate some important points:

1. It is possible to clone and express thermophilic genes in a mesophilic recipient strain providing that the appropriate vector is available.
2. For *Thermus* spp., because of the lack of good cloning vectors discussed above, genetic exchange into a mesophilic host may be the best way of exploiting this bacterium at present.
3. Such studies reveal which features are important for thermostability. In the cases described here thermostability of the cloned enzymes was similar within a mesophilic background indicating that thermostability was intrinsic and not mediated by post-transcriptional or translational modification.
4. Conversely, the use of such cloning systems should illuminate our knowledge of the different mechanisms that are able to promote thermostability on enzyme molecules.
5. This work also shows the advantages of expressing genes in a mesophilic background because they can be at least partially purified by heat denaturation of the mesophilic host's proteins.

Future prospects

SECONDARY METABOLITES

Many bacteria synthesize secondary metabolites such as antibiotics and certain extracellular enzymes and other bioactive compounds. These usually appear when growth has ceased and in some cases, notably fungi, it may be some days before the appearance of these compounds. To date relatively few thermophiles have been found that synthesize antibiotics. They include thermorubin and thermoviridin from species of *Thermoactinomyces*, thermocin from some strains of *B. stearothermophilus* and chaetomycin from a thermotolerant *Bacillus*.[22,31,84,88] Recently, granaticin synthesis has been reported to occur at 50 °C in *Streptomyces thermoviolaceus*.[45] The advantages of using thermophiles for production of secondary metabolites is that the growth and production rates are generally faster. Figure 1.2 illustrates this by comparing the synthesis of two similar antibiotics. Granaticin synthesized by *S. thermoviolaceus* at 50 °C is much more rapid than the synthesis of actinorhodin at 30 °C by *Streptomyces coelicolor*. Thus a thermophilic fermentation allows more rapid throughput by cutting down the growing period, rapid antibiotic production as well as the other advantages of operating commercial processes at elevated temperatures alluded to previously. There may be considerable future prospects in this area especially when allied to the rapid developments in cloning and sequencing of antibiotic pathways in the streptomycetes.

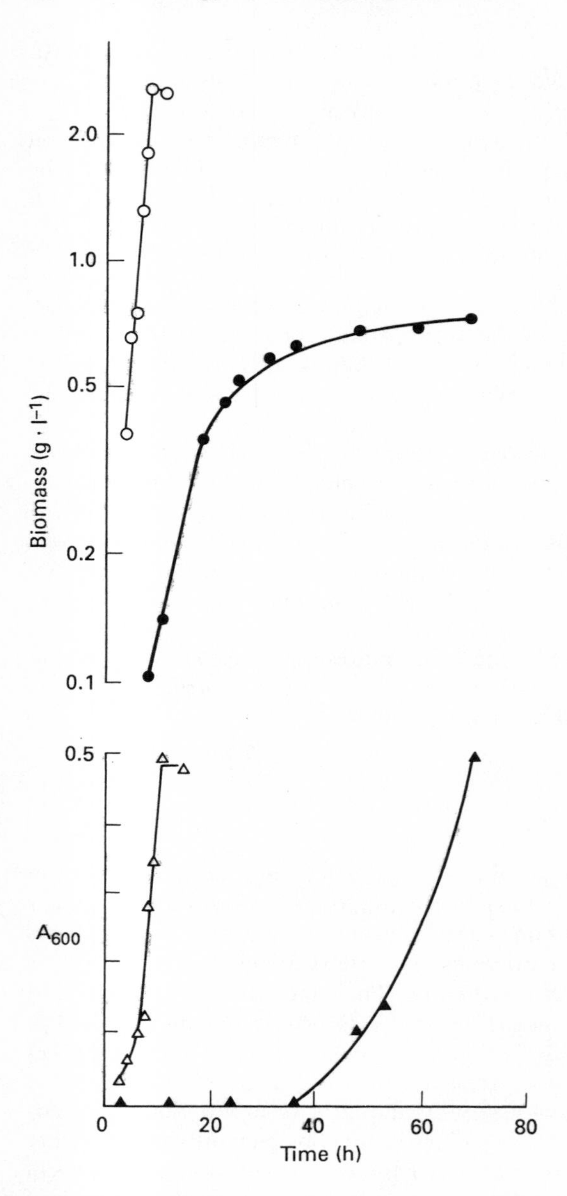

Fig. 1.2 Comparison between antibiotic production by a mesophilic streptomycete with that by a thermophilic species. Actinorhodin production (▲) was measured spectrophotometrically at 600 nm in cell free supernatants prepared at different stages of fermentation during growth (●) of *S. coelicolor* at 30 °C. Granaticin, (△) a closely related antibiotic, was monitored in the same way for *S. thermoviolaceus* grown (○) at 50 °C.

PROTEIN ENGINEERING

With the advent of more sophisticated genetic techniques, in particular site-directed mutagenesis, the possibility of engineering protein molecules for specific purposes and conditions has become a practical reality. Nowhere is this more so than for thermophilic proteins and work has concentrated on the features of enzyme molecules that can make them more thermostable. Desired properties of such proteins would include: (i) rapid catalysis; (ii) high thermostability; (iii) active and stable in organic solvents; (iv) high substrate specificity; (v) long shelf life; (vi) protease resistance. Oligonucleotide or site-directed mutagenesis requires a cloned and sequenced gene that codes for the protein of interest (from this the amino acid sequence can be calculated as well as an accurate determination of molecular weight) and a purified protein for X-ray crystallography to determine structure–function relationships. The latter aspect at present is limited.

Once the amino acid sequence is known, synthetic oligonucleotides can be synthesized whose sequence is the same as a region of the protein nucleotide sequence except that one or more nucleotides are changed. When transcribed and translated the amino acid sequence then has one or more specific and defined differences, e.g.

AGA ATT ATG AAT **TAT** TTT GAA ATG

TAT is the nucleotide triplet code for isoleucine. Preparing a synthetic oligonucleotide that reads

AGA ATT ATG AAT **TGT** TTT GAA ATG

will substitute TGT that codes for cysteine, which has the potential to introduce an additional disulphide bridge into the protein molecule as long as another free cysteine is available.

Conclusions

Studies of thermophiles have increased during the last ten years or so. The recent discovery of barothermophilic bacteria has extended the temperature range at which life can be detected to around 110 °C. Further work on these organisms will require the development of more sophisticated equipment for their cultivation. Although mesophilic organisms and their products are nearly always the choice for biotechnological processes, the advent of gene cloning will almost certainly result in the increasing use of thermophilic gene products in the future. On the whole, our knowledge of the mechanisms of thermotolerance is scant although a common feature of a number of thermophilic molecules is that they differ from their mesophilic counterparts by only small, key changes. As we understand more about protein structure and function, especially in relation to temperature, protein engineering methods will be used to tailor enzymes for specific jobs and operating conditions. Another area that seems ripe for development is that of thermophilic biodegradation processes, particularly those mediated by anaerobes. In this context it would appear that many such processes will be more efficiently achieved by consortia of different thermophiles. Knowledge of ecological interactions related to the biochemistry and physiology of such mixed populations will be of prime importance. More work on thermophilic organisms and their products should herald a better understanding of microbial life in general and the results may pose a challenge to our current ideas and dogmas.

References

1. Ahring, B.K. and Westermann, P. (1987) *Appl. Environ. Microbiol.*, **53**, 429–33.
2. Alfredsson, G., Kristjansson, J.K., Hjorleifsdottir, S. and Stetter, K.O. (1988) *J. Gen. Microbiol.*, **134**, 299–306.
3. Allais, J., Hoyos-Lopez, G., Kammoun, S. and Baratti, J.C. (1987) *Appl. Environ. Microbiol.*, **53**, 942–5.
4. Amner, W., McCarthy, A.J. and Edwards, C. (1988) *Appl. Environ. Microbiol.*, 54, 3107–22.
5. Anderson, P.J., McNeil, K.E. and Watson, K. (1988) *J. Gen. Microbiol.*, **134**, 1691–8.
6. Argos, P., Rossmann, M.G., Grau, U.M., Zuber, H., Frank, G. and Tratschin, J.D. (1979) *Biochemistry*, **18**, 5698–6703.
7. Arrhenius, S. (1927) *Zeit für Physik und Chemie der Leipzig*, **130**, 516–19.
8. Ayala, G., Gomex-Puyou, M.T., Gomez-Puyou, A. and Darszon, A. (1986) *FEBS Lett.*, **203**, 41–3.
9. Ball, A. and Edwards, C. (1986) *Arch. Microbiol.*, **145**, 347–52.
10. Ball, A. and Edwards, C. (1987) *J. Gen. Microbiol.*, **133**, 1221–6.
11. Ball, A., Edwards, C. and Jones, M.V. (1985) *FEMS Microbiol. Lett.*, **27**, 139–42.
12. Baross, J.A. and Deming, J.W. (1983) *Nature*, **303**, 423.
13. Bell, J.M., Falconer, C., Colby, J. and Williams, E. (1987) *J. Gen. Microbiol.*, **133**, 3445–56.
14. Bingham, A.H.A., Bruton, C.J. and Atkinson, T. (1979) *J. Gen. Microbiol.*, **114**, 401–8.
15. Bonjour, F. and Aragno, M. (1986) *FEMS Microbiol. Lett.*, **35**, 11–15.
16. Brock, T.D. (1978) *Thermophilic Microorganisms and Life at High Temperatures*, New York, Springer-Verlag.
17. Brock, T.D. (1986) *Thermophiles, General, Molecular and Applied Microbiology*, New York, Wiley.
18. Bryant, F.O., Wiegel, J. and Ljungdahl, L.G. (1988) *Applied Environ. Microbiol.*, **54**, 460–5.
19. Chang, S. and Cohen, S.N. (1979) *Mol. Gen. Genet.*, **168**, 111–15.
20. Chen, M. (1987) *Appl. Environ. Microbiol.*, **53**, 2414–19.
21. Croft, J.E., Love, D.R. and Bergquist, P.L. (1987) *Mol. Gen. Genet.*, **210**, 490–7.
22. Cross, T. (1968) *J. Appl. Bacteriol.*, **31**, 36.
23. Curran, M.P., Daniel, R.M., Guy, G.R. and Morgan, H.W. (1985) *Arch. Biochem. Biophys.*, **241**, 571–6.
24 Deming, J.W. (1986) *Microb. Ecol.*, **12**, 111–19.
25. Deming, J.W. and Baross, J.A. (1986) *Appl. Environ. Microbiol.*, **51**, 238–42.
26. Droffner, M.L. and Yamamoto, N. (1985) *J. Gen. Microbiol.*, **131**, 2789–94.
27. Edwards, C. and Jones, M.V. (1983). *Arch. Microbiol.*, **135**, 74–6.
28. Estival, F. and Burstein, C. (1985) *Enz. Microb. Technol.*, **7**, 29–33.
29. Fee, J.A., Kuila, D., Matter, M.W. and Yoshida, T. (1986) *Biochim. et Biophys. Acta*, **853**, 153–85.

30. Ferenci, T. and Lee, K.S. (1986) *J. Bacteriol.*, **166**, 95–9.
31. Fikes, J.D., Crabtree, B.L. and Barridge, B.D. (1983) *Can. J. Microbiol.*, **29**, 1567.
32. Fujii, M., Imanaka, T. and Aiba, S. (1982) *J. Gen. Microbiol.*, **128**, 2997–3000.
33. Fujii, M., Takagi, M., Imanaka, T. and Aiba, S. (1983) *J. Bacteriol.*, **154**, 831–7.
34. Gacesa, P. and Hubble, J. (1987) *Enzyme Technology*, Milton Keynes. Open University Press.
35. Goodfellow, M., Lacey, J. and Todd, C. (1987) *J. Gen. Microbiol.*, **133**, 3135–49.
36. Grote, M., Dijk, J. and Reinhardt, R. (1986) *Biochim. et Biophys. Acta*, **873**, 405–13.
37. Heda, G.D. and Madigan, M.T. (1986) *J. Gen. Microbiol.*, **132**, 2469–73.
38. Henson, J.M., Smith, P.H. and White, D.C. (1985) *Appl. Environ. Microbiol.*, **50**, 1547–9.
39. Hirata, H., Negoro, S. and Okada, H. (1985) *Appl. Environ. Microbiol.*, **49**, 1547–9.
40. Hocking, J.D. and Harries, J.I. (1980) *Eur. J. Biochem.*, **108**, 567–79.
41. Hoshino, T., Ikeda, T., Narushima, H. and Tomizuka, N. (1985) *Can. J. Microbiol.*, **31**, 339–45.
42. Hudson, J.A., Morgan, H.W. and Daniel, R.M. (1986) *J. Gen. Microbiol.*, **132**, 531–54.
43. Imanaka, T., Fujii, M. and Aiba, S. (1981) *J. Bacteriol.*, **146**, 1091.
44. Imanaka, T., Shibazaki, M. and Takagi, M. (1986) *Nature*, **324**, 695–7.
45. James, P.D.A. and Edwards, C. (1988) *FEMS Microbiol. Lett.*, **52**, 1–6.
46. Jones, C.W., Brice, J.M. and Edwards, C. (1977) *Arch. Microbiol.*, **115**, 85–93.
47. Kellis, J.T., Nyberg, K., Sali, D. and Fersht, A.R. (1988) *Nature*, **333**, 784–6.
48. Koyama, Y., Hoshino, T., Tomizuka, N. and Furukawa, K. (1986) *J. Bacteriol.*, **166**, 338–40.
49. Kristjansson, J.K., Hreggvidsson, G.O. and Alfredsson, G.A. (1986) *Appl. Environ. Microbiol.*, **52**, 1313–16.
50. Kruger, B. and Meyer, O. (1984) *Arch. Microbiol.*, **139**, 402–8.
51. Kubo, M. and Imanaka, T. (1988) *J. Gen. Microbiol.*, **134**, 1883–92.
52. Kuhn, H.J., Cometta, S. and Fiechter, A. (1980) *J. Microb. Biotech.*, **10**, 303.
53. Lacey, R.W. and Chopra, I. (1974) *J. Med. Microbiol.*, **7**, 285.
54. Lacis, L.S. and Lawford, H.G. (1985) *J. Bacteriol.*, **163**, 1275–8.
55. Langworthy, T.A. and Pond, J.L. (1986), in T.D. Brock (ed.) *Thermophiles, General, Molecular and Applied Microbiology*, New York, Wiley, pp. 107–35.
56. Lauers, A.M. and Heinen, W. (1983) *Antonie van Leeuwenhoek*, **49**, 191–201.
57. Ljungdahl, L.G. (1979) *Adv. in Microbial Physiol.*, **19**, 149–243.
58. Loprasert, S., Negoro, S. and Okada, H. (1988) *J. Gen. Microbiol.*, **134**, 1971–6.
59. Love, D.R. and Streiff, M.B. (1987) *Biotechnol.*, **5**, 384–7.
60. MacMichael, G. (1988) *Microb. Ecol.*, **15**, 115–22.
61. McCarthy, A.J. (1987) *FEMS Microbiol. Revs.*, **46**, 145–63.
62. McKay, A., Quilter, J. and Jones, C.W. (1982) *Arch. Microbiol.*, **131**, 43.
63. McLinden, J., Murdock, A. and Amelunxen, R. (1986) *Biochim. et Biophys. Acta*, **871**, 207–16.
64. Miquel, P. (1888) *Annals Micrographie*, **1**, 3–10.
65. Munster, M.J., Munster, A.P., Woodrow, J.R. and Sharp, R.J. (1986) *J. Gen. Microbiol.*, **132**, 1677–83.
66. Nakasaki, K., Shoda, M. and Kubota, H. (1985) *Appl. Environ. Microbiol.*, **50**, 1526–30.
67. Ollivier, B., Smiti, N. and Garcia, J.L. (1985) *Biotech. Lett.*, **7**, 847–52.
68. Perry, J.F. (1986) *Adv. Aquatic Microbiol.*, **3**, 109–40.
69. Perutz, M.R. and Raidt, H. (1975) *Nature*, **255**, 256.
70. Priest, F.G., Goodfellow, M. and Todd, C. (1988) *J. Gen. Microbiol.*, **134**, 1847–82.
71. Querol, E. and Parrilla, A. (1987) *Enz. Microb. Technol.*, **9**, 238–44.

72. Reizer, J., Grossowicz, N. and Barenholz, Y. (1985) *Biochim. et Biophys. Acta*, **815**, 268–80.
73. Russell, N.J. (1984) *Trends Biochem. Sci.*, **9**, 108.
74. Senior, A.E. and Wise, J.G. (1983) *J. Memb. Biol.*, **73**, 105–24.
75. Sharp, R.J. and Munster, M.J. (1986), in R.A. Herbert and G.A. Codd (eds) *Microbes in Extreme Environments*, London, Academic Press, pp. 215–95.
76. Sharp, R.J., Sharp, S.I., Ahmad, A., Munster, B., Dowsett, T. and Atkinson, T. (1986) *J. Gen. Microbiol.*, **132**, 1709–22.
77. Sissons, C.H., Sharrock, K.R., Daniel, R.M. and Morgan, H.W. (1987) *Appl. Environ. Microbiol.*, **53**, 832–8.
78. Sone, N., Yoshida, M., Hirata, H. and Kagawa, Y. (1977) *J. Biol. Chem.*, **252**, 2956.
79. Sonnleitner, B. and Fiechter, A. (1983) *Trends in Biotechnol.*, **3**, 14–80.
80. Strom, P.F. (1985) *Appl. Environ. Microbiol.*, **50**, 906–13.
81. Suchardova, O., Volfova, O. and Krumphanzl, V. (1986) *Folia*, **31**, 1–7.
82. Suzuki, Y., Fujii, H., Uemura, H. and Suzuki, M. (1987) *Starch*, **39**, 17–23.
83. Takagi, M., Imanaka, T. and Aiba, S. (1985) *J. Bacteriol.*, **163**, 824–31.
84. Tauritorus, T.E. and Townsley, P.M. (1984) *App. Environ. Microbiol.*, **47**, 775.
85. Watanabe, K., Himeno, H. and Ohta, T. (1984) *J. Biochem.*, **96**, 1625–32.
86. Wedler, F.C. and Hoffman, F.M. (1974) *Biochem.*, **13**, 3215.
87. Wiegel, J. (1986), in T.D. Brock (ed.) *Thermophiles, General, Molecular and Applied Microbiology*, New York, Wiley, pp. 17–37.
88. Wiegel, J. and Ljungdahl, L.G. (1986) *CRC Crit. Rev. Biotechnol.*, **3**, 39–108.
89. Yoshida, T., Lorence, R.M., Choc, M.G., Tarr, G.E., Findling, K.L. and Fee, J.A. (1984) *J. Biol. Chem.*, **259**, 112–23.
90. Yu, E.K.C., Tan, L.U.L., Chan, K.H., Deschatelets, L. and Saddler, J.N. (1987) *Enz. Microb. Technol.*, **9**, 16–24.
91. Yutani, K., Ogasawara, K., Aoki, K., Kakuno, T. and Sugino, Y. (1984) *J. Biol. Chem.*, **259**, 14076–81.
92. Yutani, K., Ogasahara, K., Sugino, Y. and Matsushiro, A. (1977) *Nature*, **267**, 274.

CHAPTER 2

Acidophiles

W.J. Ingledew

Introduction

Acidophilic organisms have evolved to exploit naturally occurring acid environments which furnish the thermodynamic potential for cell maintenance and growth.[20,24,37] Such acid niches are often created, at least in part, by the organisms that are present, i.e. acidophiles often catalyse acidogenic processes. For example, where sulphur deposits or sulphide bearing ores are exposed to a wet oxic (oxygenated) interface, the sulphur compounds can be oxidized by chemolithotrophic bacteria to dilute sulphuric acid. These bacteria play a vital role in the sulphur cycle and in geochemistry. The most studied of these processes involves the oxidation of pyritic ores (ores containing high levels of FeS_2). The Fe(II) iron is oxidized by *Thiobacillus ferrooxidans* to Fe(III) iron. The Fe(III) can oxidize the S^{2-} or polysulphide to S^0. The S^0 can then be further oxidized to H_2SO_4, normally utilizing O_2 as the terminal oxidant. This process can be catalysed by the thiobacilli and some other organisms. The result is an acidic solution of Fe(III) iron, a strongly oxidizing liquor which can bring into solution other heavy metal ions. These phenomena are exploited commercially for the extraction of Cu(II) from copper-bearing pyritic ores such as chalcopyrite[2] and uranium.[34] The converse process, solubilization of the ore to leave an enriched concentrate as residue, has been used to assist in gold extraction.[9,36] The process also poses environmental problems, causing acid run-off from old mine workings and dumps which bear high concentrations of toxic heavy metals into streams.[14,28] A number of reviews have been published in the past decade which deal with particular aspects of acidophily and the reader is referred to these for more specific information: metabolism and bioenergetics;[20,37] physiology;[4,25] the bacterial species;[15] and aspects of mining and pollution.[3,34]

Acidophiles are organisms which thrive in relatively acid media. This means more than the ability to survive for a finite time (acid tolerant); it means the organism has the ability to grow and reproduce in an acid environment. The acidophilic organisms discussed in this chapter are prokaryotes and these can be crudely split into two broad groups, facultative and obligate acidophiles.[15,25] In some cases this distinction is more a result of the chemistry of the bacterium's respiratory processes rather than a commitment of the bacterium to one particular environment. For example, the S^0 and Fe(II) oxidizing bacterium *T. ferrooxidans* can oxidize S^0 at neutral pH (although the process is acidogenic) but cannot grow at neutral pH using Fe(II) oxidation because of the autooxidation of this substrate at neutral pH.

What pH bounds constitute the range for acidophily? A cursory consideration of the stability of even the primary structures of biomolecules tells us that there must be a lower limit on the pH range which is compatible with life. The cell can protect its vital processes with a membrane enclosing a neutral microenvironment but this protecting layer itself must be biological and must have one surface exposed to the supporting medium. This consideration is borne out by observation; a few bacteria have been reported to grow at pH values close to 1.0 (100 mM H^+), but none lower.[15] In these cases, and in nearly all others of acidophily, the acidity is due to H_2SO_4; most other acids cannot support an acidophilic lifestyle. Setting the upper limit of the pH range which constitutes the bounds within which an organism must thrive in order to be defined as an acidophile is an arbitrary exercise. Some authors have taken pH 4.0 for this upper limit. As the problems of acidophily are most acute in the extreme acidophile, I shall confine most of my specific comments to prokaryotes capable of growth at pH 2.0 or lower. I shall start, however, by outlining the habitats of these organisms, consider the chemical and physiological principles which affect growth at extremes of pH, outline relevant aspects of the metabolism of the thiobacilli and of *T. ferrooxidans* in particular and finish by considering commercial and genetic aspects. Many of the examples referred to in this review are taken from the thiobacilli as these are the most extensively studied of the acidophilic bacteria.

The habitats

A cursory survey of the literature reveals that all known extremely acid niches which can sustain life are based on SO_4^{2-} as the predominant anion; they are also all low in dissolved organic matter. This impression is supported by studies on cell growth and viability; only SO_4^{2-} or its close analogue SeO_4^{2-} can sustain these.[10,41,42] There are only a few types of appropriate acidic environments but they are extensive in their range. They are generally associated with pyrite or S^0 deposits and include mining regions and geothermal areas.[6] In addition some waterlogged low lying acid soils (e.g. pine-barren areas) contain sulphur-cycling microenvironments. These niches are primarily suited to the chemolithotrophic S^0 and Fe(II) iron oxidizers. The latter niche also involves SO_4^{2-} reducing bacteria. Some heterotrophic acidophiles can also be detected in most of these niches.[15]

In the S^0 springs oxygen serves as an oxidant and the process is acidogenic; the nett reaction can be written

$$S^0 + 1\tfrac{1}{2}O_2 + H_2O \longrightarrow 2H^+ + SO_4^{2-}$$

In the oxidation of pyrite both Fe(II) and the polysulphide S_2^{2-} are oxidized; the overall reaction can be written

$$2FeS_2 + 7\tfrac{1}{2}O_2 + H_2O \longrightarrow 2H^+ + 4SO_4^{2-} + 2Fe(III)$$

These reactions are expressed in a much simplified form but show that these oxidative processes are protonogenic.

In certain soils a cycling of sulphur occurs involving the sulphur-oxidizing bacteria and SO_4^{2-}-reducing bacteria. These latter bacteria can be acidophiles although the chemistry of their redox processes is proton consuming. In the absence of oxygen these bacteria utilize SO_4^{2-} as a respiratory oxidant, the final product is H_2S (Postgate in ref. 37). In some environments this can diffuse away until it meets an oxic interface where it is acidogenically reoxidized to SO_4^{2-}. This is a serious problem in some oil industries where the acidogenic process facilitates corrosion in pipes and other structures (Jorgensen in ref. 37). In other environments the sulphide may be trapped as metal (usually Fe(II)) sulphides forming a hard-pan layer in the soil which restricts drainage. This pyritic layer can 'creep' down as the oxic interface slowly advances; this is found in acidic waterlogged soils often associated with pine-barrens.

One acidic niche that contains organic matter but which is not usually considered in the context of acidophily is the stomach. The stomach is approximately pH 2.0, at its lowest, this pH being achieved by HCl. A pH of 2.0 is compatible with the growth of acidophiles when the medium is SO_4^{2-}-based. However, the contents of the acid stomach are usually sterile. The germicidal action of gastric juice is considered to be due to the HCl. The achlorohydric stomach, whatever the cause, is usually heavily colonized with bacteria. The significant difference, as far as medium for bacterial growth is concerned, between the stomach as a niche and other acidic niches is that the dominating anion in the stomach is Cl^-.

The organisms

SPECIES

As stated above the acidic niches are primarily occupied by the chemolithotrophic autotrophs which are S^0 and Fe(II) iron oxidizers. Heterotrophic acidophiles can be found; these must exist on the organic matter fixed by the chemolithotrophs or released from fossilized deposits.[15] The chemolithotrophs cover a number of genera, *Thiobacillus*, *Sulpholobus* and *Leptospirillum* and most probably others. Recent studies indicate that some of the strains provisionally grouped under these headings are much more heterogeneous, than initially suspected.[15] The bacteria specifically considered in this article are listed in Table 2.1. The thiobacilli constitute a major group of acidophiles (there are also neutrophilic thiobacilli),

Table 2.1 Examples of types of acidophiles

Chemolithotrophs		
Fe(II) oxidizers		
L. ferrooxidans	Mesophile	Pyritic environments
Fe(II) and S^0 oxidizers		
T. ferrooxidans	Mesophile	Pyritic environments
S. acidocaldarius	Thermophile	Geothermal areas
S^0 oxidizers		
Thiobacillus sp.	Mesophile*	
Thiomicrospora	*	
Heterotrophs		
A. coyptum	Mesophile	Associated with *T. ferrooxidans* culture
Th. acidophilum	Mesophile	Coal refuse
B. acidocaldarius	Thermophile	S^0 hot springs

*There are a large number of *Thiobacilli* and other sulphur bacteria; these are extensively reviewed in ref. 37.

and they are mesophiles. *T. ferrooxidans* is the best known and understood but not the only Fe(II) oxidizer found in these environments; another is *Leptospirillum ferrooxidans*. This latter bacterium can attack pyrite by oxidizing the Fe(II) to Fe(III) but is incapable of oxidizing S^0 directly. In conjunction with *T. organoparus* extensive oxidation of the pyrite or chalcopyrite becomes possible.[5] In some geothermal areas (hot springs) thermophiles are favoured by the prevailing conditions. *Sulpholobus acidocaldarius* is a chemolithotroph which has a temperature optimum of 70 °C and a pH optimum of 2–3. It can oxidize both Fe(II) and sulphur compounds and derives its name from its unusual lobe-shape and its ability to oxidize S^0.[5]

Isolates of both mesophilic and thermophilic chemolithotrophs are often found to be contaminated by heterotrophic acidophiles. These were not previously readily detected because the cells do not normally grow in acidified rich media. However, it is now understood that at acid pH moderate concentrations of organic acids can be very deleterious to cells (see later). Acidophilic heterotrophs can grow on dilute organic matter.[14,15] Some acidophilic chemolithotrophs are facultative heterotrophs. The organic matter required for heterotrophy can be released from viable chemolithotrophs, from the breakdown of dead cells or from other environmental sources (coal, oil, other vegetation and even from the atmosphere). For reasons which are not understood some chemolithotrophs release low concentrations of pyruvate into the medium. A mesophilic acidophilic heterotroph which has been found contaminating many isolates of *T. ferrooxidans* is *Acidiphilium coyptum*. *Bacillus acidocaldarius* is an acidophilic heterotroph which is found in hot sulphur springs at between 45 and 70 °C, over a pH range of 2–6. *Thermoplasma acidophilum* is a heterotrophic acidophilic thermophile which was isolated from a coal refuse pile and grows optimally at between pH 1 and 2. This last bacterium lacks a cell wall and has been tentatively classified as a mycoplasma.[15,26]

Acidotolerance

CHEMICAL CONSTRAINTS OF GROWTH AT EXTREMES OF pH

As stated previously, there are constraints on the range of pH compatible with life. Most biochemical processes and structures are very sensitive to pH and function only close to neutrality. Further, the required tertiary and quaternary structures of proteins can be irreversibly destroyed by exposure to mildly acid or alkaline pH. The actual pH stability of a protein is a specific feature, but it has been reported that, if in an acidophile the cytoplasmic pH reaches approximately 5.0 the cell is non-viable and cannot recover.[1,4,25,33] The enzymes and metabolic processes found in acidophiles and alkalophiles are similar to those found in neutrophiles and do not have improved pH stability. Thus acidophiles and alkalophiles must maintain their cytoplasmic pH close to neutrality or die. The cytoplasmic components of a cell can be maintained at a different pH to the surrounding medium because there is an osmotic barrier between the two phases. Thus one face of the cytoplasmic membrane, and any other structure outside it (e.g. cell wall and periplasm in Gram-negative bacteria), are exposed to the external pH. In some cases specially adapted enzymes do function in this environment (e.g. some of the unique respiratory reactions of *T. ferrooxidans* and their associated enzymes). Even the primary structures of these exposed proteins, lipids and structural carbohydrates have limits to their pH stability and the peptide bonds, esters and glycosidic linkages are hydrolysed at extremes of pH. The cell cannot survive at a pH where the rate of destruction of these components is greater than the cells' ability to replace them.

Extremes of pH will also affect the mineral content of the surrounding medium and this may require some adaptation on the part of the organism. For instance, acidophiles have to tolerate high concentrations of heavy metal cations; the solubility of Fe(III) at pH 2.0 is approximately 200 mM. Conversely, at pH 10 the free concentration of Fe(III) is approximately 10^{-24} M, thus an alkalophile will require a good Fe(III) scavenging system.

PHYSIOLOGICAL CONSTRAINTS

Before considering the physiological constraints imposed upon an acidophile it is necessary to recap a few points on the principles of the theory of chemiosmosis (for a more comprehensive description of this theory see ref. 35). It is generally accepted that in cells obtaining energy for growth oxidatively, the mechanism for the coupling of the redox chemistry of respiration to the synthesis of ATP is chemiosmotic. The theory of chemiosmosis was developed by Mitchell in the late fifties and early sixties. An outline of the process is illustrated in Fig. 2.1a. Protons are pumped out of the cell by the respiratory process, represented in the diagram as a box. The process requires a closed vesicle so that the movement of protons establishes a proton-electrical potential (Δp) across the membrane. This potential is used to drive the protons back into the cell, through the ATP synthetase, or through other energy requiring processes. In moving a proton from the cytoplasm to the exterior two trans-membrane parameters are altered. A concentration difference of protons is established; this is normally expressed in units of pH

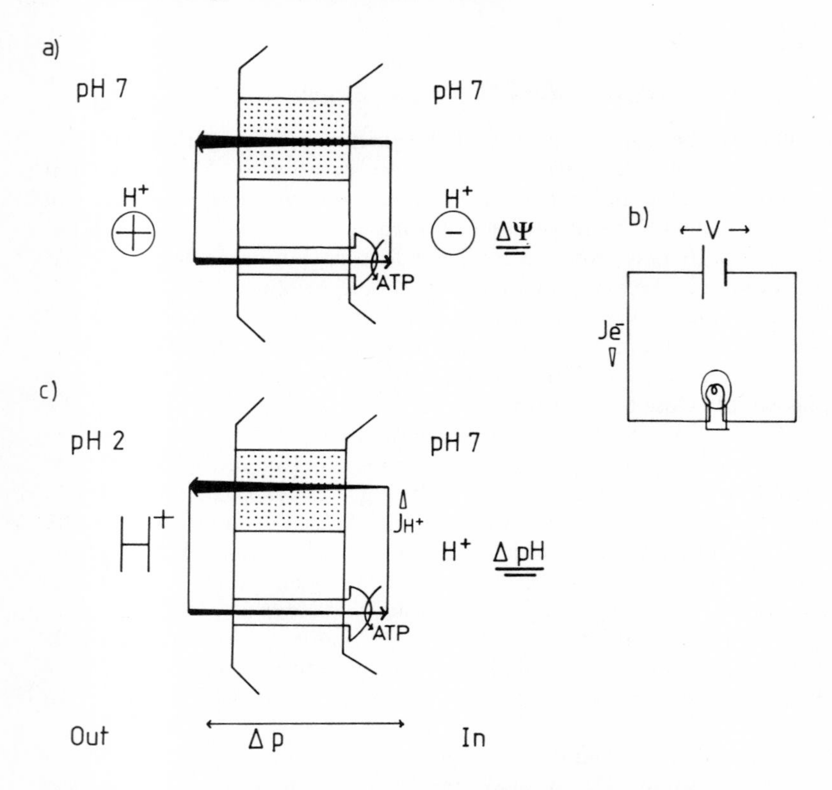

Fig. 2.1 Outline of the chemiosmotic mechanism. (a) The chemiosmotic circuit in neutrophiles. The trans-membrane pH difference is small, the **ΔΨ** is large and relatively negative inside. The proton translocating respiratory chain is shown as a shaded box, the protons re-enter the cell via the ATP synthetase. (b) An electrical analogy of the proton circuit, the battery takes the place of the respiratory chain. The lamp takes the place of the ATP synthetase. (c) The chemiosmotic circuit in an acidophile. The trans-membrane pH difference is large and the **ΔΨ** correspondingly small.

(ΔpH). In moving a proton from one side of a closed membrane to the other a charge is also moved, this creates a trans-membrane electrical potential (**ΔΨ**). Both chemical and electrical potentials are thus established by the movement of protons, and conversely they influence the movement of protons. These forces can be summated to give the nett force acting on a proton. This combined force is called the proton electro-chemical potential or proton motive force (Δp), and its two component forces contribute; $\Delta p = \Delta\Psi - 59\Delta pH$. The ATP synthetase harnesses the potential in the 'downhill' return of protons to make ATP. The respiratory process which establishes a Δp can be thought of as a battery in an electrical analogy (Fig. 2.1b). Like the electrical circuit, the proton circuit can be made, open or short-circuited. The proton circuit is made in the presence of ADP plus phosphate; it is an open circuit in their absence and it can be short-circuited by making the membrane permeable to protons (uncoupling). The Δp and the proton current required to make ATP have been calculated on thermodynamic bases and measured directly. In bacterial cells three protons and a Δp of approximately 240 mV appears to be the most likely balance. This can sustain a theoretical maximum ΔG for ATP hydrolysis:

$$\Delta G = -mF\Delta p[-3(H^+) \times 96\,000\,\mathrm{J/equivalent/V} \times 0.24\,\mathrm{V}]$$
$$= -69.22\,\mathrm{kJ/mole}$$

where m is the number of protons translocated per ATP synthesized and F is the Joule equivalent of the Faraday.

The respiratory process in acidophiles is similar to that in neutrophiles (Fig. 2.1c) except that the ΔpH is very large so that the **ΔΨ** has to be correspondingly smaller. Bacterial cells attempt to maintain their cytoplasmic pH close to neutrality. In Fig. 2.2 the values of the pH_{in}, **ΔΨ** and Δp in actively respiring cells are shown as a function of external pH. Only a few data points are included in Fig. 2.2 for clarity; these are from the acidophile *T. ferrooxidans* (half symbols), from rat liver mitochondria (solid symbols) and the alkalophile *Bacillus alcalophilus* (open symbols).[1,4,25,35] Approximate bands for the values of these parameters are shown; these only pertain to cells growing by oxidative phosphorylation, i.e. those maintaining a Δp of approximately 240 mV. If both the Δp (□–□) and pH_{in} (◇–◇) are to be maintained at 240 mV and at neutrality respectively, then the expected **ΔΨ** (○–○) will increase linearly throughout the pH range. What measurements there are show that the expected relationships hold for acidophiles and neutrophiles but they break down at alkaline pH. From published data even the expected homeostasis of the cytoplasmic pH appears to be a problem in alkalophiles. With homeostasis of the cytoplasmic pH in acidophiles and neutrophiles the transcytoplasmic ΔpH is fixed by the external pH (ΔpH = X − 7, where $X = pH_{out}$). Further, with the relationship between Δp, **ΔΨ** and ΔpH (Δp = **ΔΨ** − 59ΔpH), **ΔΨ** has to vary other pH_{out} conversely with 59ΔpH for the same value of Δp. As pH_{in} is not a variable in acidophiles and neutrophiles, it is **ΔΨ** which has to respond to the energy state of the cell.

Comments have been made as to the 'enigma' of energy conservation in acidophiles. This is misleading; the ΔpH is in the correct direction for the cell and the **ΔΨ** compensates; the system is just an altered balance of the two forces as compared with the situation in a neutrophile. In this context, the problems for chemiosmosis occur in alkalophiles not acidophiles. In the former the ΔpH is of the correct polarity to assist driving protons through the ATP synthetase; in the latter the ΔpH is of the opposing polarity. Thus, if the mechanisms are the same, the **ΔΨ** would have to be very large to give the required Δp, as illustrated on the right of Fig. 2.2, top (projected) line. In alkalophiles the necessary large **ΔΨ** has not been measured. It is also noteworthy that alkalophiles shown to grow by oxidative phosphorylation do not extend to equivalent extremes of pH as do the acidophiles. Additional ion pumps have been implicated in alkalophiles and this topic is reviewed in detail elsewhere in this volume.

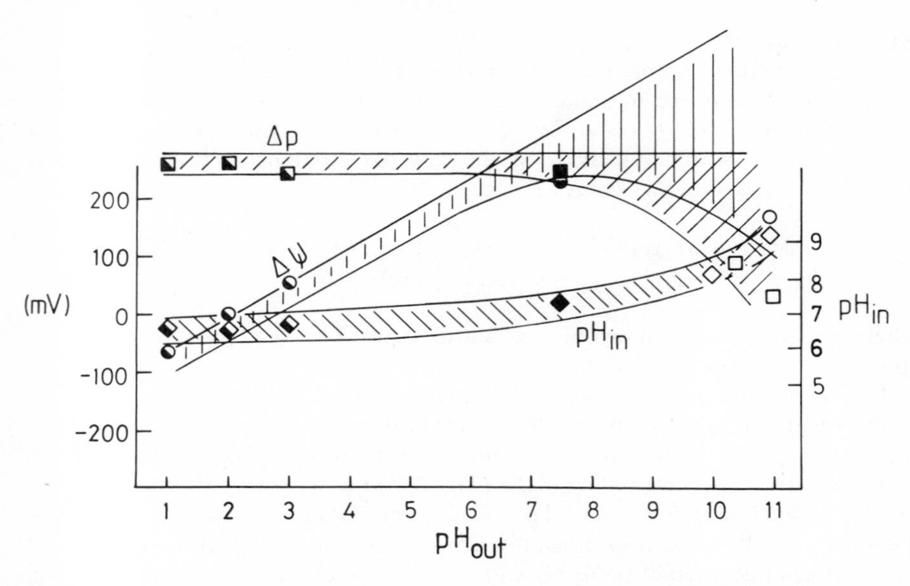

Fig. 2.2 The relationship between pH_{out} and the parameters $\Delta\Psi$, Δp and pH_{in}. Δp (□–□) and $\Delta\Psi$ (○–○) in mV on LHS axis, and pH_{in} (◇–◇) in pH units on RHS axis, are plotted as a function of pH_{out}. The halved symbols are data obtained from *T. ferrooxidans*, the solid symbols are data from mitochondria and the open symbols from the alkalophile *Bacillus alcalophilus*.

In considering the maintenance of the pH difference across the cytoplasmic membrane there are two extreme cases to consider, a 'de-energized' condition in which the cell does not input energy, and an 'energized' condition in which the cell actively maintains trans-cytoplasmic gradients, usually via respiratory linked processes. These two conditions are those in which Δp is approximately 240 mV and where $\Delta p = 0$. The fully 'de-energized' condition is one which would not be expected in a viable cell; cell reserves would normally maintain a resting state with partial energization of trans-membrane processes. The fully de-energized state can be obtained by addition of an uncoupler such as 2,4-dinitrophenol (DNP); these agents allow proton permeability of the membrane. In this condition the trans-membrane ΔpH is not abolished; it is Δp which is zero; experiments show that a ΔpH of 4.0 units can be maintained in the presence of an uncoupler, although $\Delta p = 0$.[1,8] Thus there must be an equal and opposite force acting on the proton; this is the $\Delta\Psi$. In *T. ferrooxidans*, in the presence of an uncoupler or the respiratory poison azide, the ΔpH is largely maintained but the $\Delta\Psi$ offsets this and as a consequence is large and negative. Similar results have been obtained for *Thermoplasma acidophilum* and *B. acidocaldarius*.[33]

As Fig. 2.2 illustrates, both the ΔpH and the $\Delta\Psi$ (energized or de-energized) are of different magnitude (or even polarity) in an acidophile with pH_{out} at between 1 and 2 compared with a neutrophile. This has important consequences for the acidophile in terms of its tolerances to various chemicals.

THE CONSEQUENCES OF A LARGE ΔpH, ACIDIC OUTSIDE

An acidophile in relatively acid media, if viable, will maintain its cytoplasmic pH near neutrality, hence there will exist across the cytoplasmic membrane a ΔpH. This is a large concentration difference of protons (a chemical potential) and it imposes the problems that this concentration difference can be used for the non-specific accumulation of weak acids. As a consequence acidophiles are very sensitive to simple organic acids as these can be accumulated many-fold into the cytoplasm. This may explain beliefs that true chemolithotrophs not only could not grow on organic matter but were poisoned by it. The principles behind this phenomenon are illustrated in Fig. 2.3. The ratio of the anionic to the acid form of an acid is determined by the Henderson–Hasslebach relation; $pH = pK_a + \log A^-/HA$. If we take pyruvate as an example ($pK_a = 2.5$), then at an external pH of 2.0 the anionic and protonated forms will be present at a ratio of 0.24: 0.76. With a simple organic acid such as pyruvate it has been borne out by many observations that the anionic form is not membrane permeable and the acidic form is. The consequences of HA being permeable and A^- not permeable are shown in Fig. 2.3. At equilibrium the concentration of HA on each side of the membrane is equal as no force can affect the distribution of this neutral species. On the inside the HA that has entered experiences a different pH, thus the ratio of A^- to HA on each side of the membrane will be different although the concentration of HA will be the same. Given $pH_{out} = 2.0$ and $pH_{in} = 7.0$, then the equilibrium distribution of the combination of both forms ($HA + A^-$) is 1 : 24 033, as calculated in Fig. 2.3. In practice, if the amount of weak acid added is low, the cytoplasmic pH can be maintained, but, if the concentration of the acid is significant, then the internal pH will tend to fall in response to the entry and dissociation of the acid.[1] This is a self-limiting process because as the pH falls the tendency to accumulate the anion declines. This phenomenon explains the toxicity of simple organic acids to acidophiles. The lower part of Fig. 2.3 shows a graph in which the theoretical accumulation ratios $(HA + A^-)_{in}/(HA + A^-)_{out}$ of monobasic acids are plotted as a function of pH_{out}. It has been assumed in these calculations that pH_{in} is maintained at 7.0. The influence of the pK_a of the acid on its accumulation into the cytoplasm is illustrated. The heavy line marked A represents the asymptote for monobasic acids ($pK_a < 1$) and the line marked B the asymptote for dibasic acids (pK_{a1}, $pK_{a_2} < 1$). The relationships shown in Fig 2.3 can be refined to the expression[34]

$$pH_{in} = pK_a + \log_{10}[A_{in}/A_{out}(1 + 10^{pH_{out} - K_a}) - 1]$$

for a monobasic acid. This can be extended to calculate the accumulation ratios for polycarboxylic acids:

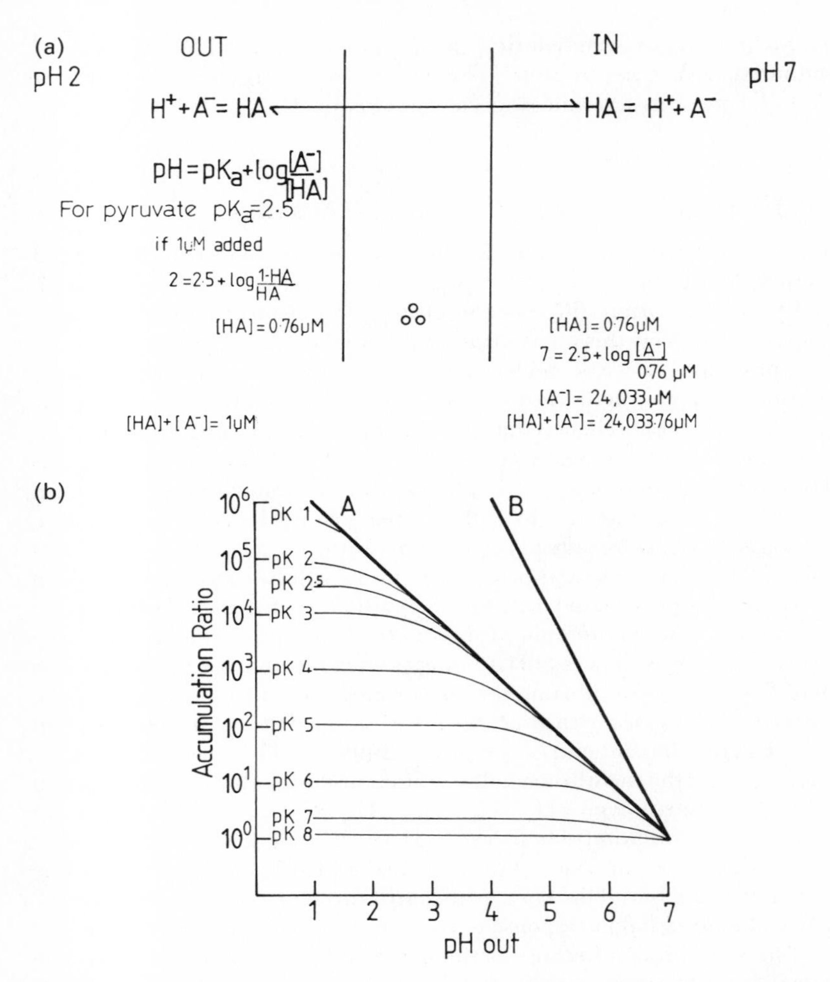

Fig. 2.3 The distribution of weak acids across the cytoplasmic membrane of an acidophile. (a) The principles, based on the Henderson–Hasslebach equation, which govern the distribution of weak acids across a membrane are shown. (b) The theoretical accumulation ratio of monobasic acids is shown as a function of their pK_a and pH_{out}, assuming pH_{in} of 7.0. The line A is the asymptote for monobasic acids. The line B is the asymptote for dibasic acids.

Divalent:

$$A^{2-}_{in}/A^{2-}_{out} + HA^{-}_{in}/HA^{-}_{out} + H_2A_{in}/H_2A_{out} =$$

$$\frac{10^{(pH_I - pK_{a2})}\,10^{(pH_I - pK_{a1})}}{10^{(pH_O - pK_{a2})}\,10^{(pH_O - pK_{a1})}} + \frac{10^{(pH_I - pK_{a1})}}{10^{(pH_O - pK_{a1})}} + 1$$

This relationship expands progressively.

Trivalent gives the additional term

$$\frac{10^{(pH_I - pK_{a3})}\,10^{(pH_I - pK_{a2})}\,10^{(pH_I - pK_{a1})}}{10^{(pH_O - pK_{a3})}\,10^{(pH_O - pK_{a2})}\,10^{(pH_O - pK_{a1})}}$$

This polynomial series simplifies if $pK_{a1} = pK_{a2} = pK_{a3} = pK_{a4}$, etc. Accumulation $= X^4 + X^3 + X^2 + X + 1$, where X is given by $10^{(pH_I - pK_a)}/10^{(pH_O - pK_a)}$. In order to be accumulated in this way there must be sufficient of the fully protonated form present in the bulk phase to equilibrate (which requires that the pK_as and pH_{out} are not too far apart). In addition the protonated form must be membrane permeable. A requirement for non-specific permeability of simple compounds is normally that it carries no charge, hence amino acids such as glutamate are not accumulated and are not toxic because the positively charged amino group renders them impermeant. A protonated but very strongly dipolar species may also be impermeant. If the A^- form is also permeant then an uncoupling cycle is formed, short-circuiting Δp. The process illustrated in Fig. 2.3 is utilized, using low concentrations of acid, to measure the internal pH of the cells. With pH_{out} and pK_a being known and the accumulation of the acid being determined experimentally, the pH_{in} can be calculated.

The consequences of the presence of a range of organic acids on the pH_{in}, $\Delta\Psi$ and the activities of the acidophile *T. ferrooxidans* have been examined in detail.[1] A range of monobasic acids were chosen and it was shown that acetate (pK_a 4.9), propionate (pK_a 4.75), lactate (pK_a 3.1), chloroacetate (pK_a 2.86) and pyruvate (pK_a 2.5) distributed according to the ΔpH and their distribution was not influenced by the $\Delta\Psi$. The ability of the acids to acidify the cytoplasm when present at millimolar concentrations and their toxicity to respiration was in approximate agreement with their predicted accumulation from their respective pK_as (Fig. 2.3).

THE ROLE AND CONSEQUENCES OF THE MEMBRANE POTENTIAL

Measurement of the $\Delta\Psi$ in extreme acidophiles has shown that this parameter is relatively small in the energized state and large and positive inside in the de-energized state ($pH_{out} = 2.0$). The $\Delta\Psi$ obtained in cells of *T. ferrooxidans* oxidizing Fe(II) is shown as a function of pH_{out} in Fig. 2.4A. The open symbols are the experimentally determined values;[1,8] the upper set of values were obtained in energized cells, the lower set in de-energized (uncoupled) cells. In these experiments the internal pH fell a little below pH 7. The solid symbols show values for $\Delta\Psi$ after adjustment for a pH_{in} of 7; this is done so that the curves (a and b) are normalized for $pH_{in} = 7$. Thus this theoretical limit on $\Delta\Psi$ can be plotted as shown in Fig. 2.4(A), holding the pH_{in} at pH 7.0. The theoretical curve a gives the expected relationship of $\Delta\Psi$ to pH_{out} where Δp is maintained at 236 mV and pH_{in} at 7.0. In the absence of energization the $\Delta\Psi$ will be equal and opposite to the ΔpH. In practice such values are only obtained in the presence of uncoupler ($\Delta p = 0$); resting cells have a residual Δp and respiring cells a Δp of between 220 and 270 mV. The uncoupled condition imposes the other limit on the membrane potential and is indicated in curve b, where $pH_{in} = 7.0$ and $\Delta p = 0$. It is only in

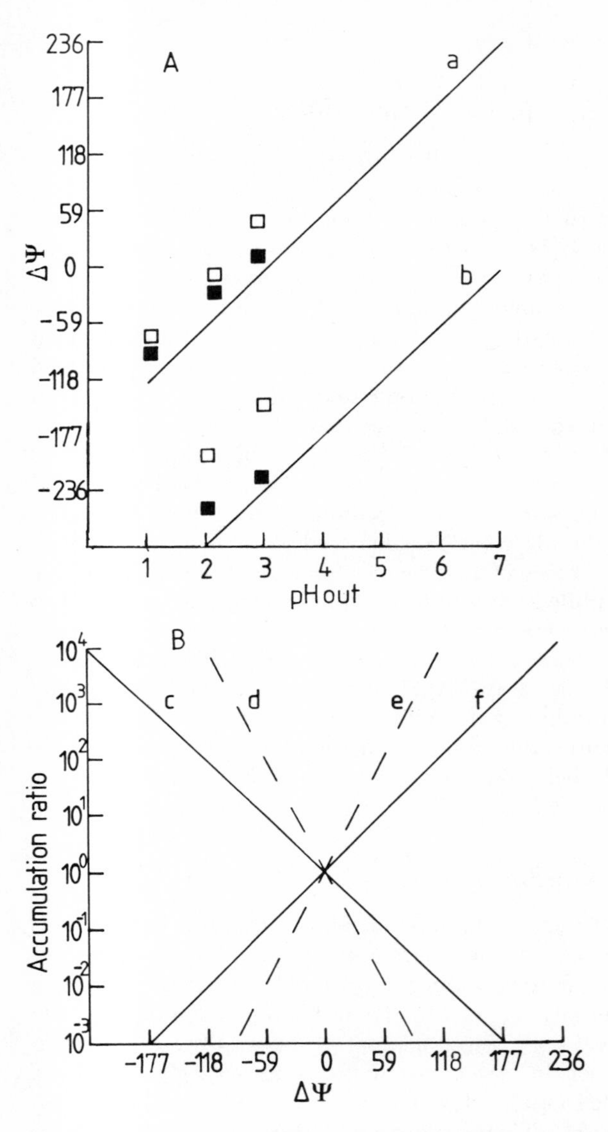

Fig. 2.4 The membrane potential and its influence on the distribution of permeable ions. (A) The relationship between **ΔΨ** and pH_{out} under energized and de-energized conditions. The **ΔΨ** was determined in *T. ferrooxidans* under respiring conditions in the absence and presence of uncouplers (□–□). The solid symbols represent the data points corrected for the fact that the pH_{in} falls slightly below pH 7 in these experiments. The curves a and b are the theoretical relationships between **ΔΨ** and pH_{out} for $\Delta p = 240$ mV and $\Delta p = 0$ when pH_{in} is 7.0. (B) The accumulation ratio of ions as a function of **ΔΨ**.

Table 2.2 Anion inhibition of Fe^{2+} respiration in *Thiobacillus ferrooxidans* (adapted from ref. 1)

	I_{50} mM addition pH 3.0		pH 0.94	
Anion	None	DNP	None	DNP
Sulphate	—	—	—	—
Chloride	10	0.5	150	1
Bromide	10	0.5	120	7
Nitrate	1	0.5	25	0.1

The I_{50} values for Fe^{2+} oxidation were determined with respect to the SO_4^{2-} controls of similar ionic strength; this did not necessarily correspond to the half maximal rate as some ions were found to stimulate respiration at low concentrations. Rates of oxygen consumption were measured by Clark oxygen electrode. DNP, where present, was at 25 μM. The sodium salts of the anions were used; the pH 3.0 buffer was 20 mM β-alanine sulphate and the basic pH 0.94 buffer was 100 mM H_2SO_4 (Data from ref. 1)

acidophiles that a large $\Delta\Psi$ positive inside can be obtained. The consequences of this are important and, like the ΔpH, expose the cell to problems and advantages not experienced by neutrophiles. A permeant ion can distribute according to the $\Delta\Psi$. This distribution is determined by the Nernst equation; e.g. for an anion

$$\Delta\Psi = 2.303RT/nF \log_{10}[X^{m-}{}_{in}]/[X^{m-}{}_{out}],$$

where R is the gas constant, T the absolute temperature, F the Faraday and n the charge of the ion; if $n=1$, then the term $2.303RT/nF$ becomes 59 mV at 30 °C. With a $\Delta\Psi$ relatively positive inside, permeant anions will tend to accumulate and permeant cations be excluded. These relationships are illustrated in Fig. 2.4B. Curve c shows the expected accumulation of a permeant monovalent cation (M^+), curve d that of a divalent cation (M^{2+}), curve e shows that of a divalent anion (A^{2-} and curve f that of a monovalent anion (A^-). Thus, with the polarity of $\Delta\Psi$ in the de-energized state in acid media, anions will tend to accumulate and cations tend to be excluded. Although the mechanism of accumulation is different from that affecting acids, it is the charged form which enters and the driving force is the $\Delta\Psi$; the end result may be accumulation and poisoning of the cytoplasm. Can this explain the unusual anion sensitivity (accumulation) of acidophiles and their unusual cation tolerance/(exclusion)? Extreme acidophiles require SO_4^{2-} as the predominant anion; only the chemically similar SeO_4^{2-} can effectively substitute.[1,27] Low concentrations of SCN^- and NO_3^- are deleterious, as are higher concentrations of Cl^-. Studies on *T. ferrooxidans* show that these ions are particularly toxic when the uncoupler DNP is present, i.e. when $\Delta p=0$ and thus $\Delta\Psi$ is large and positive inside. This is shown in Table 2.2. The effect of the anions SO_4^{2-}, Cl^-, Br^- and NO_3^- were compared. In Table 2.2 the approximate concentrations of anion required to inhibit Fe(II) oxidation by 50% in the presence and absence of uncoupler at pH 3.0 and pH 0.94 are given. SO_4^{2-} is the control anion. It can be seen from a comparison of the concentrations of anions

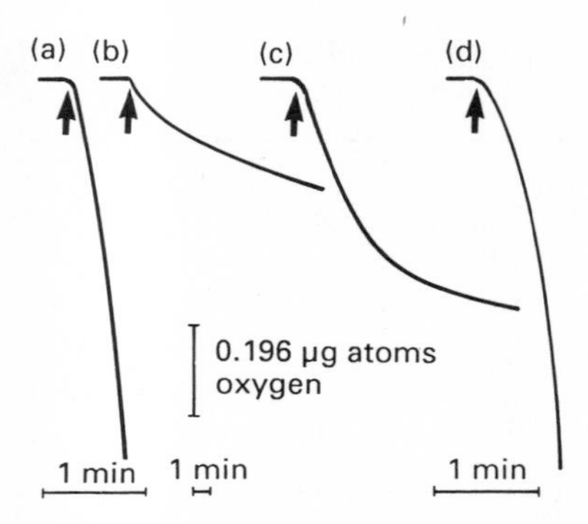

Fig. 2.5 Clark oxygen electrode traces showing the effect of Br^- ions on the oxidation of Fe(II) by *T. ferrooxidans* under different conditions (from ref. 1). (a) Control; 100 mM Na_2SO_4, 25 µM DNP in 4 ml of buffer. Substrate (2 mM $FeSO_4$) was added where indicated after 2 min preincubation of cells and buffer. (b) Substrate added last. 100 mM NaBr, 25 µM DNP. The cells and buffer were preincubated for 2 min before the addition of 2 mM $FeSO_4$. Note that the chart speed is one sixth that in the other traces. (c) Cells added last. 100 mM NaBr, 25 µM DNP and 2 mM $FeSO_4$ were placed in the vessel. Respiration was initiated by the addition of cells indicated by the arrow. (d) No uncoupler. 100 mM NaBr and cells were preincubated for 2 min prior to the addition of 2 mM $FeSO_4$ where indicated. All incubations were at 30 °C in a 4 ml vessel. The basic buffer was 20 mM *β*-alanine sulphate pH 3.0 in all incubations and the cells were added to 0.25 mg protein ml^{-1}.

required for 50% inhibition of oxidation that the presence of the uncoupler DNP greatly enhances the toxicity of the anions tested. This is because at pH 3, when the bacteria are respiring in the absence of uncoupler, the $\Delta\Psi$ will be approximately -40 mV inside (tends to electrophoretically exclude anions) whereas in the presence of uncoupler the $\Delta\Psi$ will approach $+180$ mV inside, tending to accumulate anions. At pH 0.94, respiration both in the absence and presence of uncoupler will be more sensitive to anions, because the $\Delta\Psi$ in the absence of uncoupler will be approximately $+70$ mV inside and approximately $+250$ mV in the presence of uncoupler. Sensitivity to anions is thus both a pH-dependent and an energy dependent phenomenon as a result of the effects of these on the $\Delta\Psi$. The I_{50} values in Table 2.2 were obtained from studies on the effects of the anions on Fe(II) oxidation by cells using an oxygen electrode to monitor rates. Overall the traces obtained for the effects of anions on Fe(II) oxidation are complex (Fig. 2.5) because order of addition, physiological state of the cells and length of incubation affect the rates. If cells are added to the medium containing the anion before the substrate, then the toxicity is greater, presumably because the cell has been exposed to the anion when partially de-energized. Uncoupler potentiates the inhibition. Some recovery of the oxidation rate may occur depending on the anion used, its concentration, the presence of uncoupler and the extent of preincubation. These phenomena are illustrated in the oxygen electrode traces shown in Fig. 2.5

for Fe(II) oxidation by *T. ferrooxidans* at pH 3 in the presence of the anion Br^-; trace a is the control in SO_4^{2-} media, trace b (note slower time base) shows the effect of preincubating cells with buffer, Br^-, and uncoupler before substrate addition. This is the most deleterious condition, presumably because the anion has time to accumulate inside before the onset of respiration. Trace c shows the effect of adding cells last to the incubation (in the presence of DNP); in this case the inhibition is progressive as the anion is accumulated. Trace d shows that in the absence of uncoupler Br^- is not very toxic. In experiments which measure the pH_{in} and $\Delta\Psi$ the efficacy of the range of anions tested paralleled their ability to collapse $\Delta\Psi$ and the internal pH. As with the inhibitor studies, the potency of the anions was greatly increased by the addition of DNP, thus showing that these effects are in response to the accumulation of the anion by the $\Delta\Psi$. The accumulation of NO_3^- under these conditions was measured directly by assaying the NO_3^- trapped in the cells after centrifugation through silicone oil.[1]

The above explains why a number of anions are toxic but it also raises the question as to why SO_4^{2-} appears not to be as deleterious as, say, Cl^-. As SO_4^{2-} is the control anion, it is only possible to determine it is the anion best suited to the bacterium, especially since it is found in natural habitats in which acidophiles live. It is probable that this anion can support acidophilic life for two reasons. First, as it is the dominant natural anion, the cells may have evolved active systems for removal of SO_4^{2-}. This cannot be the whole explanation as SO_4^{2-} is the best anion in the presence of uncoupler, when such porter systems could not be actively driven. Second, the anion is less permeant through membranes than the other anions. This is probably true, the sulphate anion is doubly charged and has additional polarities; even at pH 1, where the HSO_4^- form predominates, the polarity of the anion is greater and hence perhaps its non-specific permeability through lipid bilayers is less than the other common anions.

These considerations give the general principles of ion permeability and distribution in acidophiles. To a limited extent an acidophile may cope with deleterious ions by using specific porter systems. A Cl^- porter has been suggested in the acidotolerance of the acidophile *Bacillus coagulans*.[31] Such porter systems would need to be actively driven as they would be operating against the tendency of the $\Delta\Psi$ to accumulate anions, thus their use would be limited to those situations where the rate of entry of the ions was within the cells' energetic capacity to expel them.

Commercial aspects

The action of Fe(II) and S^0 oxidizers on pyritic ores produces an acid solution of Fe(III). This solution oxidizes a number of constituents in ores, the most important of these perhaps being sulphide or polysulphide to sulphur and the release of Cu(II). Other constituents of ores that are chemically oxidized may include U(IV), Sn(II) and Sb(III). Before the applications are considered, I shall give an overview of the oxidative processes.

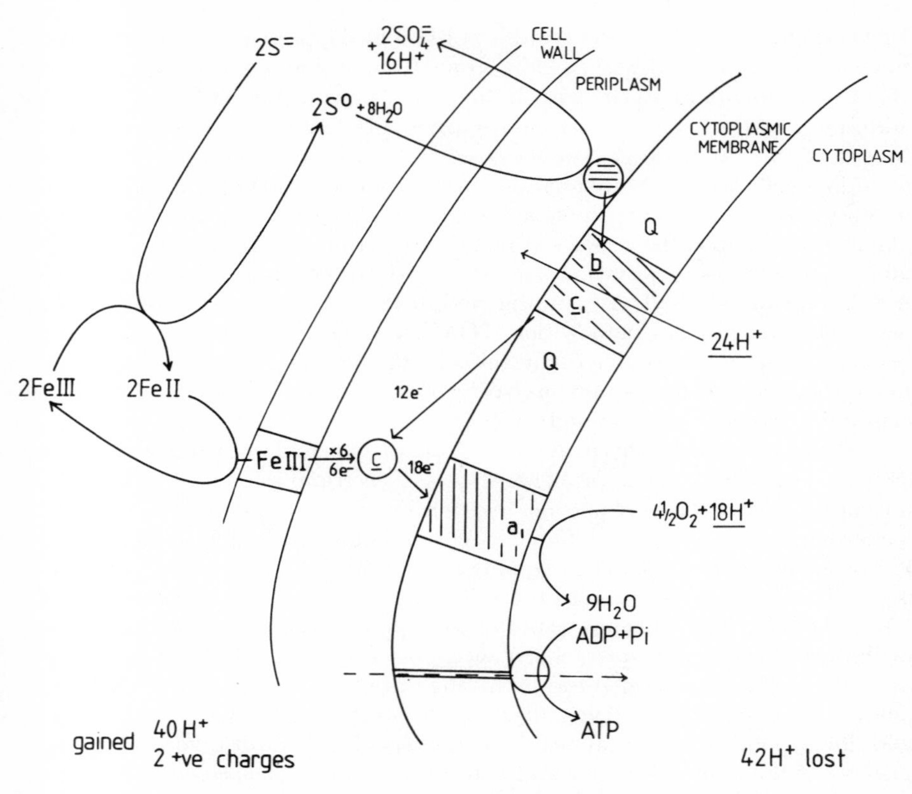

Fig. 2.6 Diagrammatic representation of *Thiobacillus ferrooxidans* oxidizing pyritic ores. In the cytoplasmic membrane two respiratory complexes are shown: the cytochrome oxidase (a_1) which is involved in all pathways to O_2 and the cytochrome bc_1 complex which is involved in S^0 oxidation. Fe(II) oxidation is envisaged as being via bound Fe(III) to cytochrome *c* (and rusticyanin, not shown) to the oxidase. S^0 oxidation is proposed to be via the bc_1 complex to cytochrome *c*. From cytochrome *c* electrons may go to O_2 or, if anaerobic, back to Fe(III). The oxidative processes establish a Δp across the cytoplasmic membrane which can be used to drive ATP synthesis.

THE CHEMISTRY AND BIOCHEMISTRY OF THE ACIDOPHILES

T. ferrooxidans, *L. ferrooxidans* and *S. acidocaldarius* can derive the energy required for growth from the oxidation of Fe(II) to Fe(III).[20,23,24] The chemical energy available from the oxidation of Fe(II) by oxygen in SO_4^{2-} media at approximately pH 2.0 is relatively small, and hence the molar growth yield is small. Yet the bacterium fixes its own CO_2[10] and, if required, its own N_2.[30] Incorporated into Fig. 2.6 is a model, proposed by Ingledew and colleagues,[20,22,23] to explain how the process of electron transport to oxygen is coupled to oxidative phosphorylation in

T. ferrooxidans. The major respiratory chain electron transfer components of *T. ferrooxidans* comprise a cytochrome a_1-type oxidase (O_2-reductase), cytochromes *c* and the copper-containing protein rusticyanin.[21] These components are organized in the cytoplasmic membrane in such a way as to couple Fe(II) oxidation to the generation of a trans-membrane Δp. In addition the cells contain a cytochrome bc_1 complex. The sided distribution of respiratory chain components has been studied by preparing spheroplasts and the electron paramagnetic probe techniques. Both methods shown that rusticyanin (a soluble blue copper protein) and a soluble cytochrome *c* are periplasmic. The cytochrome a_1 heme is buried deeply in the membrane. The generation of the Δp need not involve any trans-membrane proton translocation, merely the separation of the primary and terminal reactions by the coupling membrane as shown. *T. ferrooxidans* cells contain a polynuclear Fe(III) complex, the EPR properties of which are similar to those of phosvitin.[21,48] This polynuclear Fe(III) is bound to the phospholipid head groups of the outer membrane and elsewhere (at pH 2.0, these will be the only anionic groups in the membrane) or to a specific outer membrane protein. SO_4^{2-} could also be involved through bridging with cationic groups. What is meant by polynuclear in this context is a system of Fe(III) ions strongly interacting with each other through dipolar interactions and exchange coupling such that the behaviour of the individual ions cannot be observed. The model envisages the oxidation of the Fe(II) either in pyrite or in solution by contact with Fe(III) bound to the cell; the reducing equivalents are then transferred perhaps via rusticyanin and cytochrome *c* to the membrane-bound cytochrome a_1. The O_2 reduction reaction and consumption of protons occurs on the cytoplasmic surface. The loss of two e^- from the outside and of $2H^+$ from the cytoplasm generates a Δp for ATP synthesis.

A wider range of bacteria (all the thiobacilli and several other species) can derive energy from the oxidation of reduced sulphur compounds to (ultimately) H_2SO_4 using oxygen as oxidant. The pathways of sulphur compounds catabolism (S^{2-}, S^0, etc. through to S^{6+}) in the thiobacilli are complex and varied and have been extensively reviewed (Kuenen and Beubeker in ref. 37). In the case of sulphide from pyrite, the chemistry involves the removal of eight e^- from S^{2-}, resulting in SO_4^{2-}. In the case of S^0 oxidation six are removed to give SO_4^{2-}. Midpoint potentials for these reactions are: S^{2-} to S^0 E_{m7} -270 mV, E_{m2} 0 mV; S^0 to H_2SO_3 E_{m7} $+50$ mV, E_{m2} $+420$ mV; H_2SO_3 to H_2SO_4, E_{m7} -280 mV, E_{m2} $+163$ mV). The terminal oxidant is O_2 ($E_{m7}=800$ mV, E_{m2} $+1100$ mV) except in *T. denitrificans*, where oxides of nitrogen can be used, and in anaerobic *T. ferrooxidans*, where Fe(III) can be used as oxidant.[5,7] Unlike the exigencies suffered during growth on Fe(II), relatively large amounts of chemical energy are available for growth in inorganic sulphur compounds when O_2 is utilized as oxidant.

The first step, S^{2-} oxidation, occurs either chemically or, in the absence of Fe(III), via a sulphide oxidase. The reducing equivalents are fed into the respiratory chain and the S^0 is thought to be produced in a bound reactive form (*T. thiooxidans*, Kelly in ref. 37). The next step, S^0 oxidation is more problematical when the substrate is the native form of S^0 — the octet ring S_8^0. An initial problem is that S^0 octets are hydrophobic and insoluble and need to be opened for enzymic attack. This problem remains largely unresolved. One pathway for S^0 oxidation to

SO_3^{2-} has been demonstrated in some thiobacilli; this is the S^0 oxygenase pathway: $S^0+O_2+H_2O$ to H_2SO_3. An enzyme catalyzing this reaction has been purified 124-fold and demonstrated to be an oxygenase, containing iron–sulphur clusters (Kelly in ref. 37). This activity has been demonstrated in a number of thiobacilli, including *T. ferrooxidans* and *T. denitrificans*. The route, however, cannot be the sole one for S^0 oxidation as this process can occur at equivalent overall rates in the absence of O_2 when oxides of nitrogen (*T. denitrificans*[37]) or Fe(III) (*T. ferrooxidans*[7]) are provided as alternative respiratory oxidants. Therefore an oxidase system must also be present in these bacteria (S^0+3H_2O to $H_2SO_3+4H^++4e^-$ or S^0+4H_2O to $H_2SO_4+6H^++6e^-$). These reactions would be linked to the respiratory chain, hence to ATP synthesis. When the pathway is through the oxygenase reaction, this would not be energy yielding to the cell. Finally, the SO_3^{2-} when, and if, produced as a free intermediate is oxidized to SO_4^{2-}. Two pathways have been described: the adenosine 5′-phosphosulphate (APS) pathway and the SO_4^{2-} oxidase pathway. SO_3^{2-} will also react directly with Fe(III), if present. The APS pathway

$$SO_3^{2-}+AMP \longrightarrow APS+2e^-\ (E_m -60\ \mathrm{mV}),$$
$$APS+PO_4^{3-} \longrightarrow ADP+SO_4^{2-}$$

leads to substrate level phosphorylation as well as the possibility of oxidative phosphorylation. The alternative pathway is lower potential ($H_2SO_3+H_2O \longrightarrow H_2SO_4+2H^++2e^-$ (E_{m7} -280 mV) but gives no substrate level ATP synthesis.

Fe(II)-grown *T. ferrooxidans* retain a S^0-oxidizing system which can utilize either O_2 or Fe(III) as terminal oxidant. Both these systems are coupled to the synthesis of ATP via oxidative phosphorylation and can support cell growth. The reducing equivalents enter the respiratory chain in the region of the cytochrome bc_1 complex.[7] A model for pyrite oxidation which incorporates S^0 oxidation is shown in Fig. 2.6. The reducing equivalents from the S^0 are envisaged as passing through the cytochrome bc_1 complex before reducing cytochrome *c* and entering the common final pathway to O_2. These processes lead to the loss of H^+s from the cytoplasm and the gaining of H^+s and positive charge outside. Thus a Δp is established and maintained for the synthesis of ATP.

MICROBIOLOGICAL MINING

The processes by which the chemolithotrophs and associated bacteria attack and solubilize pyritic material have been discussed above. Only sulphidic ores are appropriate to this sort of approach and it is essential that the ore does not contain too much carbonate or other neutralizing agent to counteract the acidity or too much of a substance toxic to the bacteria. Many sulphidic ores and purified metal sulphides have been tested for bacterial attack and release of the appropriate metal ions. *T. ferrooxidans* has been shown to release Zn^{2+} from sphalerite and Pb^{2+} from galena.[12] The molybdenum containing ore molybdenite (MoS_2) is not attacked by *T. ferrooxidans* but is successfully leached by thermophiles of the *Sulpholobus* family.[5] *T. ferrooxidans* has also been shown capable of the oxidative release of Sb^{5+} from stibnite (Sb_2S_3)[43] and Sn^{4+} from stannous sulphides.[29] The release of Ni^{2+} as a consequence of bacterial action has also been noted,[5] as has the release of Ga^{3+} and MnO_2^{2-}.[43]

The above are all potential processes which are not fully developed and have not been commercially exploited although there is a lot of commercial interest in them. Commercial exploitation has occurred in only three situations, two in which the desired product is solubilized and then extracted from the liquor and one in which the product is concentrated in the residue. The solubilization of Cu^{2+} from chalcocites[19] and chalcopyrites,[44] and the solubilization of uranium from pitchblendes[13,45–47] are used extensively. More recently gold has been obtained commercially from leaching the tailings of old workings. The gold is not solubilized, but the ore is, thus the residue is a commercially viable input to the gold refining process.[9] Although there is apparently a lot of interest in this latter process, very little has been published because of its commercial sensitivity. The processings are on a relatively small scale, possibly because of the high value product, and batch extraction in reaction vessels tends to be used. Thus a range of different conditions of pH and temperature as well as different organisms are being tested. Pilot plants to desulphurize coal have been built and the principles are similar to those used in gold extraction. The pyrite is oxidized in the particulate suspension of the coal. The process is successful in removing pyritic sulphur; unfortunately much of the sulphur in coal is due to organic sulphur and it is not solubilized by *T. ferrooxidans*.[34] Some groups are trying to develop microbial leaching procedures which will also remove the organic sulphur; it is difficult to envisage how this could be attained without also oxidizing the greater part of the organic matter.

The extraction of Cu^{2+} from low-grade ore by microbiological leaching is carried out on an enormous scale. In 1980 approximately 10% of the world's copper was extracted this way and the percentage has been increasing steadily since. The process has the advantages of using low-grade ores and having low energy costs. Dump leaching is generally used; very large dumps of crushed ore are built, often filling valleys. Some of these dumps contain four billion tons of low-grade ore. Thousands of gallons of acidified water are applied by flooding or sprinkling. This process assists the onset of bacterial action and circulates the O_2 which is vital for the bacterial action. These dumps are not inoculated; the bacteria which develop are natural populations but *T. ferrooxidans* predominates with smaller numbers of *T. thiooxidans* also present. The liquor percolates through the dump and is collected at the bottom from where it may be recycled to the top, or it may have the Cu^{2+} removed before being reapplied to the top. There are a number of different methods of Cu^{2+} extraction from the 'pregnant' liquor. Generally it is removed by solvent extraction (the solvent contains a chelator of Cu^{2+}), the organic phase is then cycled over oleum and the Cu^{2+} released from the chelator into the oleum. The copper is electro-won directly from the oleum and the organic phase bearing the chelator is recycled. Another method of copper extraction is cementation in which Fe(II) replaces the Cu^{2+} in solution.

Smaller scale operations are used for uranium extraction. The smaller scale allows the use of heap leaching (sometimes also used for copper) where the construction of the heaps and grading of the ores is more carefully carried out. This gives better percolation and oxygenation and hence more rapid ore extraction. In some of these heaps aeration systems have been installed.

One of the major problems which restricts successful exploitation of ore extraction by bioleaching is the variability of ores in their susceptibility to attack. Many ores contain elements which are toxic to some of the bacteria. Given its life style, it is not surprising that *T. ferrooxidans* is relatively resistant to heavy metal ions and there are indications that it can quite readily develop resistance to ions which initially have detrimental effects.[11,45,46] At concentrations of 10 mM Cu^{2+}, Zn^{2+}, Cd^{2+}, Cr^{3+}, and Pb^{2+}, or at higher concentrations Co^{2+} and Ni^{2+}, had little or no effect on Fe(II) oxidation or cell growth.[17,18] Sn^{2+} was partially inhibitory and Ag^{+} and Hg^{+} very inhibitory. The processes are also sensitive to uranyl ions and Th^{+} and Rb^{+}. As outlined earlier many anions are toxic; of particular importance in microbial mining are the concentrations of arsenical anions and Cl^{-}.

Genetic aspects

The genetics of chemolithotrophic acidophiles have not been extensively investigated. The genetics of *T. ferrooxidans* remain essentially uncharacterized. A number of groups have isolated plasmids from *T. ferrooxidans* and genomic libraries have been constructed.[16,25,32,39,40] Progress in characterizing the genetic system has been hampered by the difficulties in applying conventional techniques to genetic analyses. The investigations into the genetics of the acidophiles are motivated by aims to improve resistance of the bacteria to toxic ions and to improve the growth rates of the bacteria. The approach taken in the former case has been to try to isolate plasmids encoding for resistance to toxic ions. To improve cell growth attempts are being made to amplify the CO_2 fixation system and the Fe(II) oxidase system.[25,39,40].

The aims of the genetic programmes may be frustrated in practice because in many applications there will be a large wild-type population. The hopes to isolate and transfer plasmids carrying resistance to toxic ions assumes resistance can be conferred by a few enzymes.

Future prospects

As the availability of high-grade ores declines, the viability of commercial exploitation of microbiological mining of suitable ores improves, given the above stated limitations. This viability is linked to the price of energy; microbiological mining is a low-energy process; normal refining of most ores is energy intensive. Thus, if energy becomes more expensive and high-grade ores less readily available, commercial exploitation of microbial mining may be applied to some of the test systems mentioned above. In extracting these ores it is unlikely that the bacterial populations can be controlled but it may be of benefit to speed up the natural processes by introducing phages capable of imparting resistance to toxic substances found in the ores. The efforts in genetic engineering of these bacteria, although constrained by competition with a vast wild-type population, may give transient advantage in this situation.

The use of microbial leaching in the partial desulphurization of coal may be extensively used as pressure against acid rain increases.

References

1. Alexander, B., Leach, S. and Ingledew, W.J. (1987) *J. Gen. Microbiol.*, **133**, 1171–9.
2. Beck, J.V. (1967) *Biotech. Bioeng.*, **9**, 487–97.
3. Blundell, T. (1979) *Trends in Biochem. Sci.*, **4**, 77–80.
4. Booth, I. R. (1985) *Microbiological Reviews*, **49**, 359–78.
5. Brierley, C.L. (1982) *Sci. Amer.*, **247** (2), 44–53.
6. Brock, T.D. (1978) *Thermophilic Microorganisms and Life at High Temperatures*, New York, Springer-Verlag.
7. Corbet, C. and Ingledew, W.J. (1987) *FEMS Microbiology Lett.*, **41**, 1–6.
8. Cox, J.C., Nicholls, D.G. and Ingledew, W.J. (1979) *Biochemical Journal*, **178**, 195–200.
9. Curtin, M.E. (1983) *Biotechnol.*, **1**, 228–35.
10. Gale, N.L. and Beck, J.V. (1967) *J. Bact.*, **94**, 1054–9.
11. Golomzik, A.I. and Ivanov, V.I. (1965) *Mikrobiologiya*, **34**, 465–8.
12. Gormely, L.S., Duncan, D.W., Dranion, R.M.R. and Pinder, K.L. (1975) *Biotechnol. Bioeng.*, **17**, 31–49.
13. Guay, R., Silver, M. and Torma, A.E. (1977) *Biotechnol. Bioeng.*, **19**, 727–40.
14. Harrison, A.P. (1978) *Appl. Environ. Microbiol.*, **36**, 861–8.
15. Harrison, A.P. (1984) *Ann. Rev. Microbiol.*, **38**, 265–92.
16. Holmes, D.S., Lobos, J.H., Bopp, L.H. and Welch, J.C. (1983) *J. Bact.*, **157**, 324–6.
17. Imai, K., Tsuyoshi, S. and Tano, T. (1972) *Proc. 4th International Fermentation Symp., Ferment. Techn. Today*, **5**, 521–6.
18. Imai, K., Sugio, T., Yasuhara, T. and Tano, T. (1972) *Nippon Kogyo Kaishi*, **88**, 879.
19. Imai, K., Sakaguchi, H., Sugio, T. and Tano, T. (1973) *J. Ferment. Technol.*, **51**, 865–70.
20. Ingledew, W.J. (1982) *Biochim. Biophys. Acta*, **683**, 89–117.
21. Ingledew, W.J. and Cobley, J.G. (1980) *Biochim. Biophys. Acta*, **590**, 141–58.
22. Ingledew, W.J., Cox, J.C. and Halling, P.J. (1977) *FEMS Microbiol. Lett.*, **2**, 193–7.
23. Ingledew, W.J. and Houston, A. (1986) *Biotechnol. and Appl. Biochem.*, **8**, 242–8.
24. Kelly, D.P. (1971) *Ann. Rev. Microbiol.*, **38**, 265–92.
25. Krulwich, T.A. and Guffanti, A.A. (1983) *Advances in Microbial Physiology*, **24**, 173–214.
26. Kulpa, C.F., Roskey, M.T. and Mjoli, N. (1986) *Biotechnol. and Appl. Biochem.*, **8**, 330–41.
27. Lazaroff, N. (1977) *J. Gen. Microbiol.*, **101**, 85–91.
28. Leathen, W.W., Braly, S.A. and McIntyre, L.D. (1953) *Appl. Microbiol.*, **1**, 65–8.
29. Lewis, A.J. and Miller, J.D.A. (1977) *Canad. J. Microbiol.*, **23**, 319–24.
30. MacIntosh, M.E. (1978) *J. Gen. Microbiol.*, **105**, 215–18.
31. McLaggan, D. and Matin, A. (1986) *EBEC Reports*, Vol. 4, Amsterdam, IUB–IUPAB, p. 417.
32. Martin, P.A.W., Dugan, P.R. and Touvinen, O.H. (1981) *Canad. J. Microbiol.*, **27**, 850–7.
33. Michels, M. and Bakker, E.P. (1985) *J. Bact.*, **161**, 231–7.

34. Murr, L.E., Torma, A.E. and Brierley, J.A. (1978) *Metallurgical Applications of Bacterial Leaching and Related Microbiological Phenomena*, New York, Academic Press, pp. 491–520.
35. Nicholls, D.G. (1982) *Bioenergetics: An introduction to chemiosmotic theory*, London, Academic Press.
36. Pol'kin, S.I., Yiedina, I.N., Panoin, V., Kostyiekova, U.N. and Koroslyschovskii, N.B. (1980) *Gidrometallurgiya Zoltata*, **M**, 67–71.
37. Postgate, J.R. and Kelly, D.P. (eds) (1982) *Sulphur Bacteria*, Cambridge University Press.
38. Webb, J. (1975), in C.A. McAuliffe (ed.) *Techniques and Topics in Bioinorganic Chemistry*, New York, Halstead, pp. 271–306.
39. Rawlings, D.E., Pretorius, I.M. and Woods, D.R. (1984) *J. Biotechn.*, **1**, 129–34.
40. Rawlings, D.E., Pretorius, I.M. and Woods, D.R. (1984) *J. Bact.*, **158**, 737–8.
41. Razzell, W.E. and Trussell, P.C. (1963) *J. Bact.*, **85**, 595–603.
42. Schnaitman, C.A., Korczynski, M.D. and Lundgren, D.G. (1969) *J. Bact.*, **99**, 552–7.
43. Torma, A.E. (1978) *Canad. J. Microbiol.*, **24**, 888–91.
44. Torma, A.E. and Gabra, G.G. (1977) *Antonie van Leeuwenhoek*, **43**, 1–6.
45. Torma, A.E. and Itzkovitch, I.J. (1976) *Appl. Environ. Microbiol.*, **32**, 102–80.
46. Tuovinen, O.H. and Kelly, D.P. (1974) *Arch. Microbiol.*, **95**, 165–80.
47. Tuovinen, O.H. and Kelly, D.P. (1974) *Arch. Microbiol.*, **98**, 78–83.
48. Tuovinen, O.H., Niemella, S.I. and Gyllenberg, H.G. (1971) *Biotechnol. Bioeng.*, **13**, 517–27.

CHAPTER 3

Alkalophiles

R.G. Kroll

Introduction

Both eukaryotic and prokaryotic microorganisms capable of growth at several values of pH above neutrality have long been known. Having been often considered as oddities they have received little attention until the last decade when the alkalophilic bacilli have been intensively studied.[33] Their existence and potential is now more widely recognized, not only for several practical applications, but also for providing potential insights into aspects of the molecular mechanisms of energy transduction, particularly at the level of membrane transport processes and the relationship between the structure and function of enzymes.

This has been due to the recognition of the central role of protons (and hence pH) in microbial bioenergetics,[75] the application of the techniques of molecular biology to these organisms[32] and the realization that many extremophilic bacteria (particularly thermophiles and halophiles) can be representatives of a phylogenetically distinct kingdom.[95] This has reorientated our views on the diversity of microbial life.

The environments frequented by extremophiles are certainly remote from standard laboratory conditions and it is axiomatic that the ability of free-living organisms to survive and/or grow at high (or low) values of pH must confer distinct survival and reproductive advantages in certain circumstances. However, it will become clear that to date knowledge of alkalophilic life is almost entirely confined to certain members of the genus *Bacillus*. Our knowledge of other alkalophiles is at best patchy and apologies are due for concentrating almost entirely on these organisms throughout. Perhaps, in another decade, other alkalophiles will have received as much investigation. For the present we are forced to make the very dangerous assumption that some of the features that characterize the alkalophilic bacilli are relevant to other alkalophiles.

Habitats

There are very few naturally occurring alkaline environments on Earth; the most notable being several soda lakes and deserts, whose ecology and chemistry have been studied in detail,[20] where pH values of about 10 are common, some dilute alkaline springs and desert soils, and soil containing decaying proteins.[19,33,65] There are also a few manmade alkaline environments in wastes from food or other industrial processes.[17,20] In natural environments it is generally found that sodium carbonate is the major source of alkalinity and this is usually included in media designed for the isolation of alkalophilic organisms, although potassium carbonate, sodium bicarbonate, sodium borate, sodium orthophosphate and even sodium hydroxide can be used.[19,33] Indeed, the media, materials, growth conditions and methods for the isolation and cultivation of alkalophiles have been well described[19,33] and will not be reiterated. It should, however, be noted that some alkalophiles, especially from soda lakes, are also halophiles and sodium chloride must be included to isolate these organisms.[19]

It is not certain exactly what makes alkaline environments inhibitory to many forms of life. That pH is a fundamental parameter affecting the growth of microorganisms has long been known[65] but it must be an oversimplification that the effects of pH are due to concentration of protons *per se*; indirect and direct effects must combine gradually to make a variety of vital processes more difficult to operate as the pH gets further from optimal values. Proton concentration will affect metabolic redox reactions and the ionic state and therefore the availability of many metabolites and ions, e.g. Fe^{2+}, Ca^{2+}, Mg^{2+}, which may become insoluble and precipitate, particularly at alkaline pH.[65] Many cell constituents, i.e. proteins, nucleic acids, lipids, carbohydrates, etc. which contribute to cell functions such as enzyme catalysed metabolism, protein synthesis, nucleic acid replication, must have at least part of their structures exposed to an aqueous environment (not just water). Many of these components contain charged groups (e.g. JH_4^+, COO^-, PO_4^{3-}) which will ionize and complex with ions and water molecules. Thus pH will have a profound effect on the biological and chemical activity of these compounds. For instance, enzyme structure will be directly affected by pH by alterations in the folding of the molecule and/or dissociation of monomers at extremes of pH. This will affect the access of both substrates and coenzymes to the active site and will directly affect the activity of the enzyme. Furthermore, the circulation of protons (or hydroxide equivalents) across cytoplasmic, chloroplast and mitochondrial membranes is an essential feature of several cellular processes.[75] Therefore any excess or deficit of protons in the surroundings must make these processes more difficult to operate. Indeed, it is probable that the pH at which an organism starts to lose control of transmembrane proton movements will be close to the inhibitory pH for that organism; the devotion of energy needed becomes too great or a breakdown point is reached, e.g. membrane impermeability degenerates.[59]

The organisms

DEFINITIONS

There are no precise definitions of what characterizes an alkalophilic or an alkalotolerant organism. An alkalophile can be considered to be unable to grow at pH 7 or less, have optimal growth around pH 9 and be capable of growth above pH 10.[19] Alkalotolerant organisms are capable of growth or survival at pH values in excess of 10 but have optimal growth around neutrality. Krulwich[59] describes organisms that can grow around neutrality but grow well or even best at pH 9–9.5 as alkalotolerant, and excludes many bacteria from being alkalophiles, restricting this term to organisms that exhibit little or no growth below pH 8.5. Furthermore, some alkalophiles are described as being facultative with respect to pH.[21] Indeed, from present data it is often difficult to ascribe an organism to a distinct category but here we will use the following, simpler and less strict definitions.

Alkalophile: An organism whose optimum rate of growth is observed above at least two pH units above neutrality.

Alkalotolerant: An organism capable of growth or survival at pH values in excess of 9.0 but whose optimal rate of growth is around neutrality or less.

TYPES OF ALKALOTOLERANT AND ALKALOPHILIC ORGANISMS

Many different organisms have now been isolated and it seems that there is no fundamental phylogenetic restrictions on whether certain groups of organisms can have alkalophilic or alkalotolerant members. It is probably true that only a fraction of the extant organisms have been isolated and characterized, especially when both alkalotolerant and alkalophilic bacteria can be isolated not only from alkaline environments but from soil and seawater[11,33,68] and that alkalophiles can be halophilic,[20,68] thermophilic[46] and psychrotrophic.[47] Undoubtedly new organisms will be added to the fairly exhaustive list of both alkalophilic and alkalotolerant organisms known to date (Table 3.1). It should be emphasized that the taxonomy of many of these organisms is rather limited and caution is needed. For example, many alkalophilic *Bacillus* species do not readily fit into existing taxonomic profiles[33] and some tentative identifications can turn out to be representatives of a new genus.[9,17]

Bacteria

It is clear that many bacteria are alkalotolerant, e.g. nitrate reducers, sulphate reducers, some *Rhizobium* spp., several enteric bacteria (e.g. *Aeromonas*, *Yersinia*) and enterococci.[65] However, true alkalophiles are well known in the genus *Bacillus* (e.g. *B. pasteurii*, *B. firmus*, *B. alcalophilus*) and their spores germinate optimally at alkaline pH.[33] There are many other reports of different alkalophilic bacteria but of particular interest are the cyanobacteria, which can give rise to massive blooms in soda lakes,[19,20,65] have optimal growth at around pH 9–10, with some species reported to grow at pH 13 (Table 3.1). These are valuable food sources for birds like flamingoes, and even for humans in some parts of the world.[20] Other organisms of interest are the alkalophilic photosythetic bacteria such as *Ectothiorhodospira*, a

Table 3.1a Alkalophilic (alk-ph) and alkalotolerant (alk-t) bacteria

	pH_{min}	pH_{max}	pH_{opt}	*Reference*		*Comments*
PHOTOSYNTHETIC BACTERIA						
Cyanobacteria	4	9–10	6–8	65	alk-t	Often abundant in alkaline environments
Gloothece linaris			10	65	alk-ph	
Microcystis aeruginosa			10	65	alk-ph	
Arthrospira plantensis		>11		65	alk-ph?	
Plectonema nostocorum		13!		65	alk-ph?	Highest recorded pH for growth
Spirulina sp.	8	>11		19, 20	alk-ph	
Anabaenopsis sp.		>11		19, 20	alk-t	
Synechococcus leopoliensis	7.0	9.3		20, 74	alk-t	Na^+ requirement for alkalotolerance demonstrated
Other photosynthetic bacteria						
Ectothiorhodospira sp.	6	11.0	8–9.5	19, 25	alk-ph	Halophilic photoautotroph
Halobacterium sp.			9–10	20	alk-ph	Extreme halophile
Chromatium			8.5	20	alk-t	
Thiospirrilum			8.5	20	alk-t	
Rhodopseudomonas (and others)			8.5	20	alk-t	
Natronococcus				20	alk-ph	Inhibited by Mg^{2+}
Natronobacterium				20	alk-ph	Inhibited by Mg^{2+}
NON-PHOTOSYNTHETIC BACTERIA						
Flavobacterium sp.	8	11.4	9–10	33, 65	alk-pH	
Nitrosomonas spp.		10.7–13		33, 65	alk-t	Survival only at alkaline pH — no growth
Nitrobacter spp.		10.7–13		33, 65	alk-t	Survival only at alkaline pH — no growth
Rhizobium		10–11		65	alk-t	
Many enteric bacteria	*c*4	9–10	6–8	65	alk-t	Including several potentially pathogenic organisms
e.g. *Yersinia enterocolitica*				87	alk-t	Used as basis for selective isolation
Aeromonas sp.				64	alk-t	
Enterococci				65	alk-t	
Unidentified bacterium		>10	8	46	alk-t	Sporogenic thermophile (opt temp = 50 °C)
Pseudomonas spp.	7.3	10.6	9	68	alk-ph	Isolated from sea water
Vibrio spp., e.g. *alginolyticus*	10	10.6	≤9	11, 68	alk-ph and alk-t	Isolated from sea water
Corynebacterium sps.				33, 47, 53	alk-ph	Isolated from soil, some psychrotrophic
Micrococcus sp.	7	11.0	9–10	33	alk-ph	
Exiguobacterium aurantiacum				9, 17	alk-ph	Isolated from potato process waste
Arthrobacter sp.	6	12.5	9	33	alk-ph	Can alkalify suspending medium

Table 3.1a cont.

	pH_{min}	pH_{max}	pH_{opt}	*Reference*		*Comments*
Unidentified sp.	5	9.5	7.5–8.0	91	alk-t	Cellulytic, thermophilic (75–85 °C) spore-forming anaerobe
Unidentified Gram +ve rod	7.4	11.4	10.0	47	alk-ph	Psychrotroph isolated from soil
Unidentified Gram − ve rod	7.0	11.0	9.5–10	47	alk-ph	Psychrotroph isolated from soil
Agrobacterium sp.				19	alk-t	
Methanohalophilus zhilinae				70	alk-ph	Halophilic methylotroph grows in 10M ammonium hydroxide but only in the presence of soil particles. No extant cultures
Bacillus pasteurii	*c*9	11		65	alk-ph	Requires amonium chloride for growth
B. alcalophilus, firmus	>8	11.5	9–10	33, 65	alk-ph	
B. sphaericus, pantothenticus	5	11		65	alk-t?	
B. rotans, circulans	5	11		65	alk-t?	
Actinomycetales						
Nocardiopsis dassonvillei subsp. *prasina*	7.0	11.0	9–10	76	alk-ph	
Streptomyces (1206 strains) (e.g. *Nocardia asteroides*)		>10.3		86	alk-t?	151 strains showed antimicrobial activities
Streptomyces (6 strains)	6.0	11.5	*c*7.0	72, 73	alk-t	
Streptomyces (56 strains)	7.0		9–9.5	72, 73	alk-ph	

Table 3.1b Alkalophilic (alk-ph) and alkalotolerant (alk-t) eukaryotic organisms

	pH_{min}	pH_{max}	pH_{opt}	*Reference*		*Comments*
Yeasts						
Exophila alcaliphila	6	10		33	alk-t	
Candida pseudotropicalis	2.3	8.8		33	not even alk-t?	
Saccharomyces fragilis	2.4	9.05		33	not even alk-t?	
Fungi						
1151 strains from 15 genera (*Absidia, Chaetomium, Penicillium, Botrytis, Ceraspora, Cephalosporium* (2 sp.), *Cladiosporium, Colletotridium, Curvularia, Fusarium* (3 sp.), *Gibberella, Helminthosporum, Manilia, Pyriculoria, Sclerotium*)		>10.3		86	alk-t?	148 strains showed antimicrobial activity
Cyrysosporium (2 sp.)				35	alk-ph?	Keratinolytic isolates from birds' nests
Aphanoascus fulvescens				33		
Penicillium variabile	*c*2.0	11.0		65	alk-t	Many fungi have wide pH tolerance

Table 3.1b cont.

	pH_{min}	pH_{max}	pH_{opt}	Reference	Comments
Fusarium (2 sp.)	*c*2.0	11.0		65	alk-t
Green algae and diatoms					
Chlorella sp.		10.0		65	alk-t
Nitzchia, Navicula, Cyclotella				20	alk-t?
Protozoa					
Euglena gracilis		11.0		65	alk-t
Crustacea					
Chydorus	3.0	10.0		65	alk-t
Zooplankton					
Brachtonus plicatilis (rotifer)				20	alk-t?
Paradiaptomus africanus (copepod)				20	alk-t?

Halobacterium sp., and the newly described haloalkalophilic archaebacteria, *Natronococcus* and *Natronobacterium*[20] and a haloakalophilic methylotroph.[70] Truly alkalophilic *Pseudomonas* and *Vibrio* spp. (and alkalotolerant spp.), a *Micrococcus* sp., an *Arthrobacter* and an unidentified strict anaerobe (Table 3.1) have also been isolated, as have psychrotrophic Gram-positive rods and cocci, coryneforms, and a Gram-negative rod.[47] Whether these are unrepresentative exceptions is not certain. Numerous alkalophilic and alkalotolerant Actinomycetales, particularly streptomycetes, have also been described[72] and this is of obvious commercial interest as streptomycetes have proved such valuable sources of antibiotics.

Fungi and yeasts

Many fungi have wide tolerance to pH and can grow between pH 2 and 11.[65] These organisms are probably alkalotolerant although alkalophilic fungi associated with birds' nests have been described[35] (the pH optima and ranges were not stated and we must conclude that these are alkalotolerant fungi). However, an alkalophilic fungus, *Paecilomyces lilacimus*, has been described[86] which grew in the pH range 2.5–11.0 and well between pH 7.5 and 9.0. Whether this can be truly described as an alkalophilic fungus is doubtful.

Alkalophilic yeasts have not been described but there is one report[33] of an alkalotolerant *Exophila* yeast. Some other yeasts (Table 3.1) appear to be able to grow around pH 9.0 but whether they can be described as alkalotolerant is marginal. A characteristic of yeasts is that they tend to be found in acidic conditions (e.g. fruit juices, etc.) and microbiological media designed for their isolation can be acidic to preclude the growth of bacteria.

Other organisms

It is worthy of note that alkalotolerant green algae and diatoms, protozoa, zooplankton, a crustacean and even fish have been described.[20,65]

Alkalotolerance

Before exploring the possible mechanisms and adaptations of bacteria to high pH, it is appropriate first to emphasize the fundamental importance of protons in several membrane transport and bioenergetic processes in bacteria, and why, in some cases, alkalophiles pose problems for our understanding of these processes. The exact mechanisms of these processes, e.g. how the protons are extruded by cells, whether they all appear in the bulk extracellular aqueous phase and the stoichiometries of many ion-linked systems and proton pumps are the subjects of continuing debate which are described in detail elsewhere.[5,6,13,39,59,62,80,81,83]

HOMEOSTATIC CONTROL OF pH_i IN NEUTROPHILES

When methods were developed for the measurement of the proton motive force (Δp) in bacteria[5,39,80,81] it quickly became apparent that both components of the Δp were not directly interchangeable or constant and they were very dependent upon the extracellular pH (pH_o). The trans-membrane pH gradient (ΔpH) was shown (initially with *E. coli*) to be several orders of magnitude at acidic pH_o values, the intracellular pH (pH_i) being alkaline with respect to pH_o but non-existent at pH_o 7.6–7.8, and the ΔpH could even be reversed at more alkaline values of pH_o so that the pH_i was more acidic than pH_o. The value of the membrane potential ($\Delta\Psi$) was also pH_o dependent, being larger at more alkaline values of pH_o. This meant that *E. coli* apparently tries to maintain its pH_i at about pH 7.6–7.8. This ability to regulate pH_i was found in several organisms (Fig. 3.1) and obviously suggests a mechanism by which free-living organisms could tolerate changes in pH_o within certain values. The homeostatic control mechanism(s), which maintain pH_i relatively constant, allow intracellular processes such as enzyme activities to be unaffected by pH_o. For example, pH_i values for acidophiles (pH_o 1–5) of 6–7, neutrophiles (pH_o 5–8.5) of 7.2–8.0 and alkalophiles (pH_o 9–11) of 9–9.5 have been obtained (Fig. 3.1). Only a few representative organisms of each group have been studied in detail with respect to their mechanisms of pH_i regulation, thus care must be taken when considering these groups as a whole.

Before summarizing what is known about these aspects, it is worth emphasizing what characteristics might be expected from pH_i controlling systems. The initial response to any imposed stress must involve changes in the activity of the existing systems to minimize further perturbations. Only on longer time scales can changes in gene expression occur to modify the cells' behaviour.[6] For trans-membrane transport systems the two major homeostatic bacterial systems that have been characterized well are for regulation of pH_i and osmotic pressure.[6,26] Any homeostatic mechanism must have some feature(s) of feedback control since changes in activity must occur in order to achieve the aims of the regulatory response. Such changes are limited by the extracellular and intracellular availability of the solutes needed to achieve any modifications. It is also axiomatic that control of pH_i requires that the routes (of which there are many) and relative rates of proton exit and entry must be controlled, that sufficient energy is available to control and move protons, and that these transport systems are sensitive to pH_i

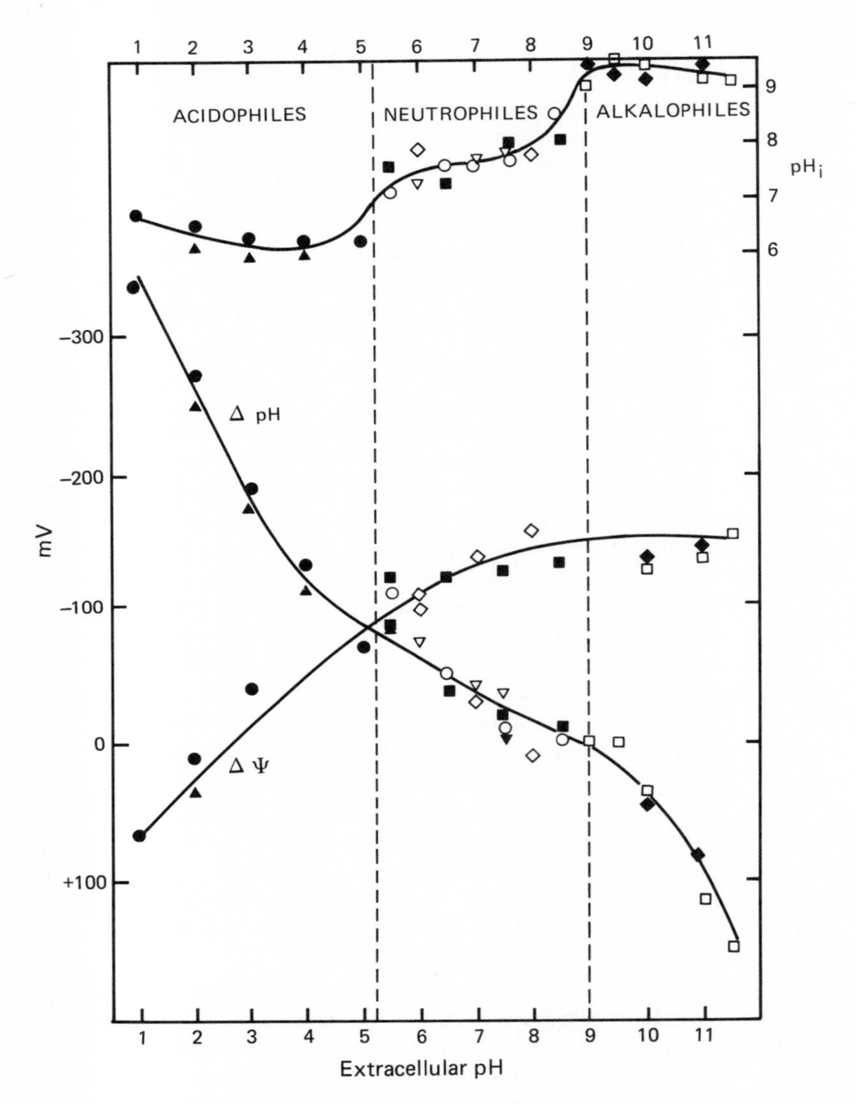

Fig. 3.1 Values of ΔpH, $\Delta\Psi$ and pH$_i$ as a function of pH$_o$. *E. coli* (◇), *Micrococcus lysodeikticus* (○), *Halobacterium halobium* (■), *Enterococcus faecalis* (▽), *Bacillus subtilis* (▼), *B. acidocaldarius* (▲), *Thiobacillus ferroxidans* (●), *B. alcalophilus* (□), *B. firmus* (◆). Reproduced from ref. 82 with permission.

(and perhaps pH$_o$). It should be emphasized that a particular problem in designing unequivocal experiments to understand these mechanisms has been the identification of the major stresses on pH$_i$ control that any particular cell has to counteract.[5]

The simplest and most widely understood physiology of pH$_i$ regulation is that of *Enterococcus* (*Streptococcus*) *faecalis*. Being non-respiring, this organism generates $\Delta\Psi$

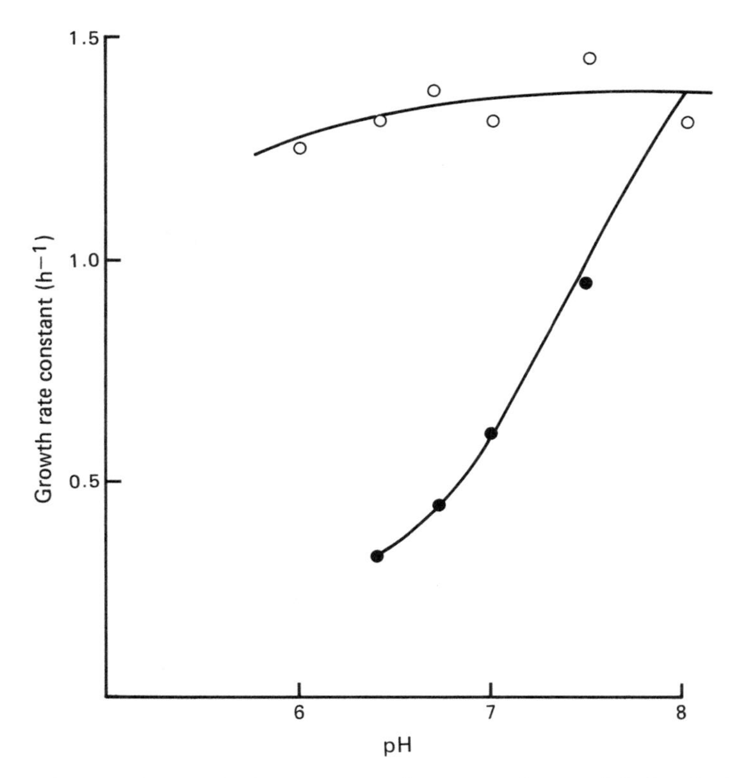

Fig. 3.2 Effect of pH on the growth rate of cultures of *E. faecalis*: (○) control; (●) when treated with gramicidin D ($4\,\mu g\,ml^{-1}$) which equilibrates the gradients of H^+, Na^+ and K^+ so that pH_o and pH_i are the same. As modified by Booth[5] and reproduced from ref. 23 with permission.

and ΔpH by the operation of the Mg^{2+}-dependent H^+-translocating ATPase which pumps H^+s out of the cell using ATP generated by substrate-level phosphorylation. With this organism the need for the maintenance of pH_i was first demonstrated.[23] Gramicidin was used to render the membrane permeable to H^+, Na^+ and K^+ so that no trans-membrane gradients of these ions existed. Providing that the extracellular K^+ and amino acid levels were kept high and Na^+ ions low, the organisms grew perfectly well when compared to the non-gramicidin treated control, but only when the pH_o (and therefore pH_i) was about 7.6. As the pH_o was lowered, growth rate decreased and eventually ceased, showing that pH_i needed to be maintained alkaline at acidic values of pH (Fig. 3.2). Indeed, the primary mode of action of weak acid food preservatives is to reduce pH_i to inhibit growth.[5]

The mechanism of pH_i regulation in this organism has been shown to be solely due to the operation of the ATPase. The major pH_i-affecting load in this organism is the intracellular acidification due to acidic fermentation products. As pH_i falls, energy is expended by the hydrolysis of ATP to pump protons out of the cell via the ATPase to raise/maintain pH_i alkaline with respect to pH_o. Interestingly, control of pH_i does not appear to be achieved at the level of direct alterations in the level of activity of this membrane transport process (as with other organisms — see later). Lowering the cytoplasmic pH of these cells stimulates the synthesis of the ATPase to increase proton pumping capacity. When cells are transferred from pH_o 7.5 to pH_o 6, an immediate fall in the pH_i from 7.7 to 6.9 occurs. The pH_i then recovers to 7.5 and this is dependent on the increased synthesis of the ATPase.[52] Whether this is the only response to fluctuations in pH_i is unclear. Thus, pH_i regulation is seen as a simple affair in this organism.

It was not thought that a mechanism for cytoplasmic acidification is necessary in this organism (although Na^+/H^+ exchange is known[36]). However, a recent report[37] has shown that carbonate is necessary in the medium for this organism to grow at alkaline pH_o where the organism is capable of maintaining pH_i acidic. Neither proton extrusion nor Na^+/H^+ exchange are necessary for this to occur and there is no evidence of a K^+/H^+ antiport. Carbonate appears to be necessary for pH_i acidification and is perhaps linked to a requirement for carbonate for fermentation by this organism at high pH. However, the requirement for carbonate is not specific, a non-metabolizable buffer (CHES) could equally support growth and the operation of an unidentified proton pump has been suggested.[37]

Exactly how neutralophilic bacteria like *E. coli* achieve pH_i homeostasis is not clear. It should be remembered that, as with many alkalophiles, high rates of proton flux due to oxidative phosphorylation and other proton-dependent transport systems must be occurring and the cell must control the net rate of proton exit and entry in order to maintain both the levels of $\Delta\Psi$ and ΔpH (*sic* pH_i). In neutralophiles, unlike alkalophiles, Δp is maintained at approximately 200 mV over the range of pH_i for growth. As a consequence of pH_i regulation, the $\Delta\Psi$ component also appears to be regulated, increasing at alkaline values of pH_o. The precise factors controlling the magnitude of $\Delta\Psi$ are not understood. It could simply be a consequence of pH_i regulation or an increase of membrane permeability to a certain ion, e.g. K^+ at acidic values of pH. However, it seems likely that the value of $\Delta\Psi$ is controlled (by dissipation of $\Delta\Psi$?) by the feedback control of an unidentified ion channel which is sensitive to Δp.[6]

What seems certain is that high rates of respiration can occur at both high and low values of pH_o without the generation of a significant alkaline interior pH_i and that the ΔpH can only be generated by the net exchange of protons for K^+ ions by the K^+ uptake systems.[5]

It has been proposed that the net proton efflux and entry (i.e. pH_i) is controlled by either or both a K^+/H^+ antiport and a Na^+/H^+ antiport which operate to acidify the cytoplasm at neutral and the more alkaline values of pH_o.[81] Mutants lacking either antiport activity and showing altered pH_o dependency have been isolated, but suffice to say that the evidence for their involvement in pH_i regulation is controversial.[5,6] Recent studies in *Vibrio alginolyticus* and *E. coli* have shown that K^+/H^+ antiport activity is regulated by pH_i.[77] Addition of a permeant amine to cause rises in pH_i at alkaline pH_o provoked K^+ efflux which slowed as pH_i was restored to its usual values. The activity of the antiport was low when pH_i was at normal values and was not affected by the high trans-membrane K^+ gradient which would favour K^+ efflux, indicating a high degree of control not related to K^+ gradients but dependent upon pH_i.

Recent genetic studies suggest that there are in fact at least three, and possibly four, K^+/H^+ antiports in *E. coli*. When the activity of two of these (*KefB* and *KefC*) were eliminated by transposon mutagenesis, pH_i regulation and K^+/H^+ antiport activity were normal.[6] Such multiple systems perhaps suggest the importance of controlling pH_i in cell survival. It should be stated that the evidence for the involvement of the Na^+/H^+ antiport in pH_i regulation is also quite strong[5,80,81,83] but the case for the role of both antiports has yet to be clarified.

The operation of such mechanisms for specifically bringing protons into the cytoplasm must be wasteful in energy terms (metabolic energy being used to pump protons out by respiration) but the ecological advantage of regulating pH_i in survival terms must outweigh this. This waste of energy would be particularly acute using the K^+/H^+ antiport to acidify the cytoplasm as energy is expended to accumulate the K^+. The large pool of K^+available to the K^+/H^+ antiporter should, however, make it thermodynamically competent to bring in large numbers of protons if necessary, particularly at alkaline pH. Using the Na^+/H^+ antiport would appear to be less wasteful, as the Na^+ gradient generated by the operation of the antiport could be used to drive the uptake of several solutes. This might also require the operation of a Na^+ entry system (see later). However, a Na^+ requirement for *E. coli* has never been demonstrated.

With *E. coli* the negative $\Delta\Psi$ will tend to attract protons intracellularly and this implies that expulsion of cytoplasmic protons is the major problem for the cell. It could be that the antiports are simply proton sinks, of which there are many others, and there is no need for the operation of acidifying mechanisms. However, the recent evidence of the pH_i-dependent activity of the K^+/H^+ antiport is against such a simple view.[77]

pHi REGULATION IN ALKALOPHILES

As a general feature of alkalophiles is to maintain pH_i more acidic with respect to pH_o, mechanisms for bringing about the controlled acidification of pH_i have been sought. The need for active mechanisms for controlling pH_i are now generally accepted, although intracellular buffering capacity could contribute to reducing fluctuations in pH_i. The cytoplasmic buffering capacity of *B. alcalophilus* and *B. firmus* have been measured and compared with that of other bacteria.[60] The cytoplasmic buffering capacity of the alkalophiles was significantly higher at higher values of pH but was very similar to that of a non-alkalophile, *B. subtilis*. High buffering capacities were also observed at low pH. However, since rapid changes in pH_i can occur on the loss of active mechanisms for pH_i regulation, internal buffering capacity cannot offer lasting protection.[60]

In several alkalophilic bacilli and *Exiguobacterium*, not only is Na^+ required for growth but it has now been established that Na^+ transport plays a particularly significant role[3,6,59,62,67] in controlling pH_i. Some of these findings can be summarized thus:

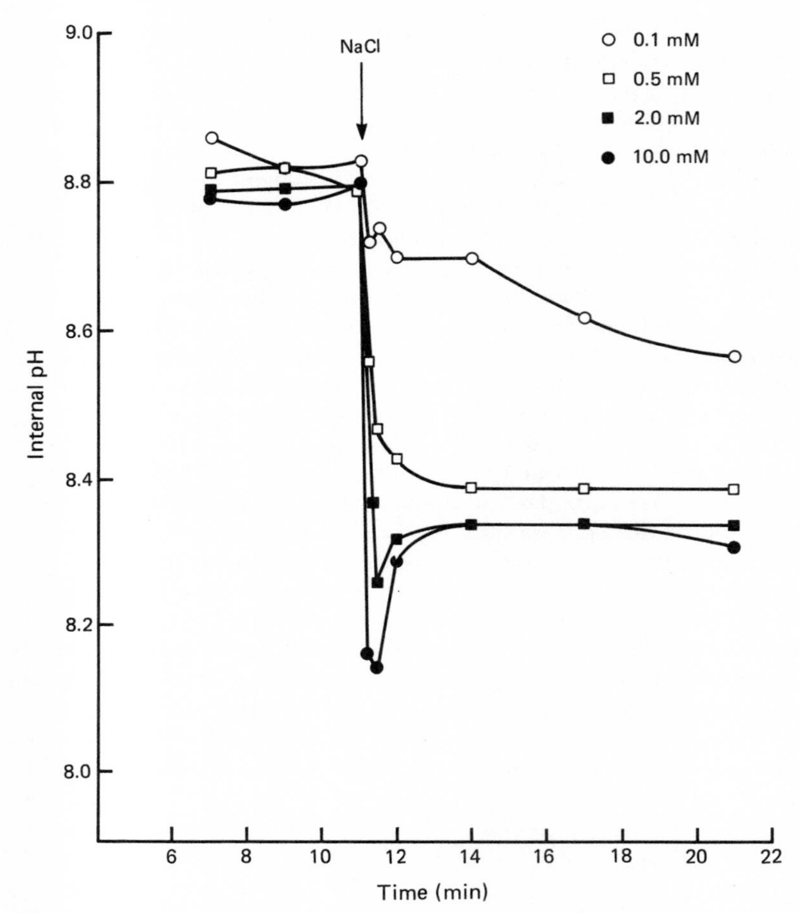

Fig. 3.3 Na^+-dependent acidification of the cytoplasm of *E. aurantiacum*. Cells were incubated for 7 min at pH 9.0 and then the pH_o was raised rapidly to pH 9.65 with KOH. After 4 min different concentrations of NaCl were added. Reproduced from ref. 67 with permission.

1. The pH_i of cells of *B. firmus* incubated in Na^+-free media increases but does not do so in the presence of Na^+, the requirement for Na^+ being greater at alkaline pH_o. Similarly, the pH_i of *E. aurantiacum* was more acidic in the presence of Na^+ and cells exhibited a controlled acidification, the rate and extent of which depended on the addition of Na^+ ions[67] (Fig. 3.3).
2. Everted membrane vesicles exhibit $\Delta\Psi$ and Na^+-dependent proton extrusion (in neutrophiles respiration would give alkalization).
3. Respiring membrane vesicles of *B. alcalophilus* and *B. firmus* require Na^+ to reverse their pH gradient and non-alkalophilic pleotropic mutants (single locus) have been isolated which have lost Na^+/H^+ antiport activity and fail to exhibit this Na^+-dependent acidification.

It is envisaged that the sodium proton antiport is primarily responsible for the acidification of the cytoplasm (Fig. 3.4). To be kinetically competent to do this requires that there is sufficient intracellular Na^+. This appears unusual as this antiport is generally assumed to result in the maintenance of low intracellular Na^+. However, there seems no reason to suppose its function could not have become adapted to function in pH_i regulation as well. The intracellular requirement for Na^+ implies that Na^+ circulation across the membrane is necessary. There are several routes for Na^+ entry into the cell, e.g. via Na^+ symport, and it has been shown that the presence of a non-metabolizable substrate for a Na^+ symport enhances Na^+-dependent pH homeostasis in *B. firmus*.[58,61] However, all these routes have other cell functions and, although the antiporter does have pH-dependent activity, it is difficult to see how it alone could directly control pH_i.

The kinetic competence of the antiport to bring protons into the cell is determined by intracellular levels of Na^+. It has been proposed that if pH_i becomes too alkaline, a pH_i-sensitive Na^+ channel brings Na^+ into the cell, allowing the activity of the antiport to increase and bring in more protons.[5] Some indirect evidence does exist for this with *Exiguobacterium*. When pH_i is acidic, fluxes of Na^+ are limited. When external Na^+ is raised, a transient increase in intracellular Na^+ is seen, followed by efflux of the cation. Moreover, the peak internal Na^+ concentration is significantly smaller when the initial pH_i is low, suggesting that under this condition either entry is prevented or exit accelerated.[67] There is also some evidence that alkalophiles must possess a mechanism by which over-acidification of the cytoplasm is prevented.[6] The exact nature of the systems involved here is not known. In conclusion, the Na^+ cycle is the major route for proton entry and aspects of the cycle are pH-sensitive but it cannot be the sole mechanism of pH_i regulation. Indeed, both neutral-sensitive and alkaline-sensitive mutants of facultative bacilli have now been isolated[55,56] confirming that both alkalization and acidification systems are necessary, probably by specific exchange of H^+ for K^+ and Na^+ at neutral and alkaline pH, respectively. These studies have also suggested that the pH-dependent effects of K^+ on $\Delta\Psi$ result from a pH-dependent change in the K^+ permeability of the membrane.

Another aspect of note is that *B. firmus* can produce variants which will grow well on media containing less Na^+ (2–3 mM) than usually required (5 mM). Most of these variants are unstable but stable isolates can occur.[63] These variants grow better than the parent strain at high pH values and have enhanced Na^+/H^+ antiport activity. This suggests a genetic mechanism whereby cells could adapt to more alkaline conditions by amplification of a gene(s) (compare with *E. faecalis*) and it was shown that there was a large increase in the amount of a 90 kD membrane protein, presumably the antiport.

The mechanisms of control of pH_i in other alkalophiles are not known. One recent study[57] suggests that the picture of pH_i regulation in alkalophiles above may not be a universal feature of alkalophilic growth. A facultatively anaerobic alkalophile was isolated which contains no cytochromes or quinones. The organism grows optimally at pH 9.5–10 and is able to maintain pH_i acidic. Remarkably, the organism was capable of maintaining pH_i acidic even in the

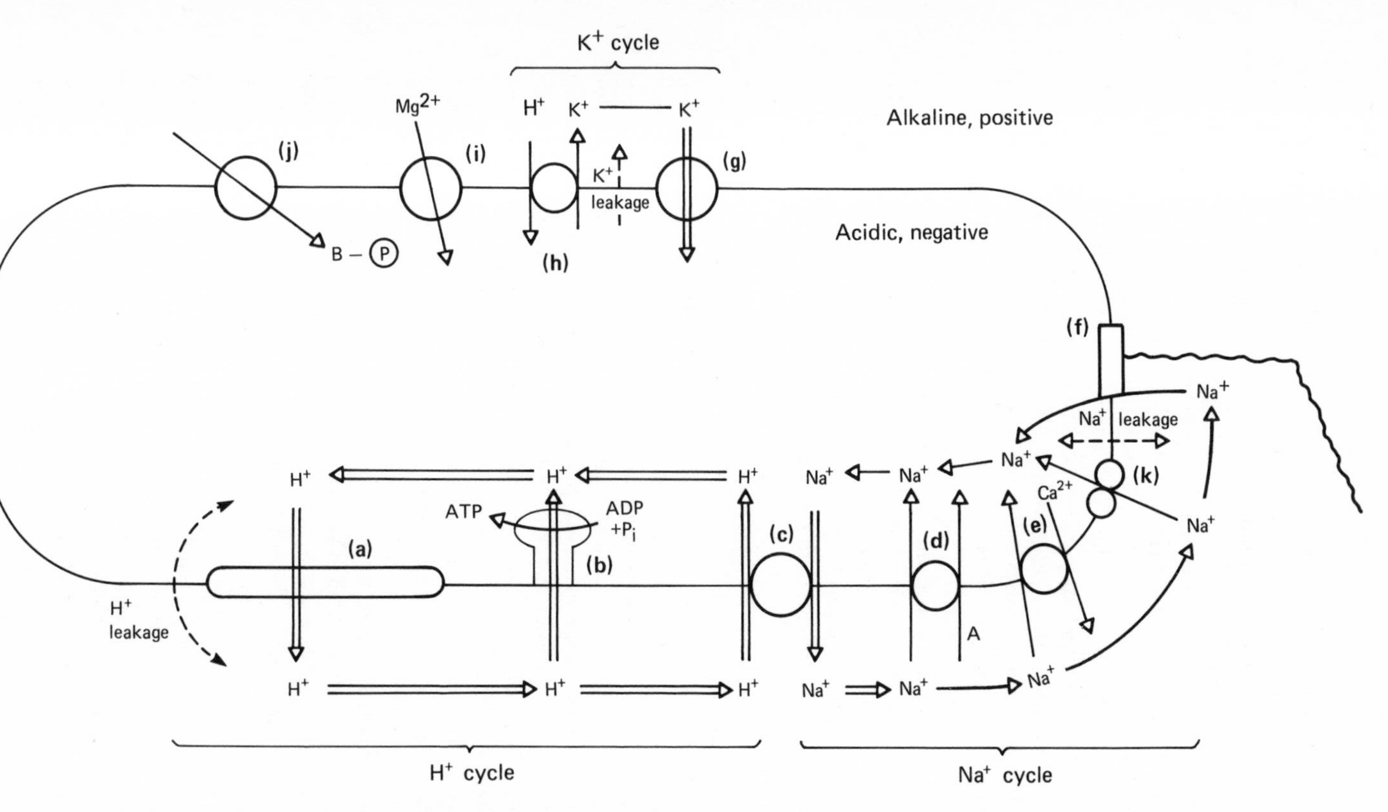

Fig. 3.4 Schematic representation of the major cation transport processes across the cytoplasmic membrane of an alkalophilic *bacillus*. (a) Respiratory chain; (b) H^+ translocating ATP synthase; (c) Na^+/H^+ antiport; (d) Na^+ symport driving uptake of compound A; (e) Na^+/Ca^{2+} antiport; (f) flagellar motor; (g) K^+ uptake system; (h) K^+/H^+ antiport; (i) Mg^{2+} uptake system; (j) phosphotransferase system. Note that no stoichiometries are implied in this diagram. The K^+ and Mg^{2+} uptake have not been characterized but may bear some resemblance to those of *E coli*. A K^+/H^+ antiport is known, as is a Ca^{2+}/Na^+. It is tempting to speculate that a K^+/Na^+ antiport exists which could be the postulated PH_i sensitive Na^+ entry system (k). The major H^+ flux must be via the ATPase and the Na^+/H^+ antiport, depending on environmental conditions.

absence of Na^+, although acidification is more efficient in its presence. This acidification might be due to Na^+/H^+ exchange energized by glycolysis, although a role for the Donnan potential has been suggested.[57]

Growth of alkalophiles

Obviously the remarkable feature of alkalophiles is that they grow optimally at alkaline values of pH (Table 3.1). An unusual aspect is that some alkalophiles exhibit more than one pH optimum.[20] For heterotrophic organisms there seems to be no explanation for this but it has been suggested, e.g. with *Ectothiorhodospira*,[20] that this may be related to the redox potential of the electron acceptor, i.e. different sulphur compounds may give different pH optima (pH 7–8.5) from organic compounds which can give more alkaline pH optima (pH 9–10). Thus some organisms can appear to be either alkalotolerant or alkalophilic depending on the growth conditions.

There are no characteristic temperature ranges for the growth of alkalophiles. Alkalophilic bacilli grow best at between 25 and 45 °C but temperature resistant isolates (up to 57 °C) and several thermophilic and even psychrotrophic alkalophiles are known.[33,47] However, it has been observed that a better tolerance to high pH is observed with some *Clostridium* isolates when the incubation temperature is lowered.[38] No precise growth factors for alkalophily have been described. Alkalophilic bacilli require growth factors as represented by other members of the genus. They will grow well in normal growth medium (provided the pH is right) although some species do require vitamins. However, one of the most characteristic features of many alkalophilic bacilli is a profound requirement for Na^+ ions. Sodium is also required for the growth and photosynthesis by an alkalotolerant cyanobacterium (*Synechococcus leopoliensis*[74]) at alkaline pH. However, a Na^+ requirement by some species, e.g. *B. alcalophilus* has not been demonstrated.[59] This organism has a more alkaline pH optimum when grown on malate than on lactose (and a higher pH for growth).

Bacillus pasteurii is unusual as it requires both alkaline conditions and ammonia. The organism grows well at pH 11 but poorly at pH 9 unless ammonium chloride is added.[33] Several strains of closely related bacteria are also capable of growth at these pH values but they do not require ammonia (e.g. *B. sphaericus*, *B. pantothenticus*, *B. rotans*). Perhaps the best known is *B. alcalophilus* which grows actively at pH 10.0 but not at pH 7.0. The requirement for ammonia by *B. pasteurii* is not understood. It may reflect a high nitrogen requirement or perhaps the lack of an ammonium transport system so that high concentrations are needed extracellularly.

Many alkalophiles can survive in media with low concentrations of divalent cations (Mg^{2+} and Ca^{2+}), normally scarce in alkaline environments, but high levels, e.g. with *Natronococcus* and *Natronobacterium*, can be toxic.[20] The characteristics of the transport systems in these organisms should be interesting to study.

Most bacteria produce either neutral or acidic metabolic end-products when grown in laboratory media. However, it is known that *E. coli* can alter its metabolism in response to pH_o actually to alter pH_o and drive it towards neutrality.[33] Under acidic conditions deaminations predominated whereas under alkaline conditions considerable acid production occurred. It is interesting to note that a similar phenomenon occurs in alkalophilic bacilli. When incubated at pH_o 12, acid is produced to bring the final pH_o to 9.[33] At more acidic values of pH_o, the bacteria increases the pH_o to again about pH_o 9, presumably by the action of proteases. However, the expression of alkaline proteases does not appear to be very pH dependent.[33]

BIOCHEMICAL ADAPTATIONS IN ALKALOPHILES

It is probable that all cellular structures outside the cytoplasmic membrane (including the outer face of the cytoplasmic membrane) are exposed directly or indirectly to the pH_o (assuming the cell envelope does not restrict the movement of protons). Thus the cell wall, outer membrane and periplasmic enzymes (in Gram-negative bacteria), flagella, pili and extracellular enzymes may be adapted, if necessary, to function at high pH_o. Some biochemical modifications could be expected and it is known, for instance, that alkalophilic bacilli grown at pH 10 have a large surface charge than when grown at pH 8.[33]

Flagella

Several alkalophilic bacilli are motile and vigorous motility is observed at alkaline pH.[59] Being largely extracellular, flagella will be exposed to pH_o and some modifications might be expected to allow them to operate at alkaline pH_o. The flagella of *B. firmus* have been shown to be very similar to that of other neutrophilic bacilli, being composed of a single protein subunit (flagellin, molecular weight 40 kD) and having a similar amino acid composition except that fewer basic amino acids are present. This may render the flagella more stable at alkaline pH_o. It is also clear that unlike other organisms, such as *E. coli*, where the flagellar motor is driven by the influx of protons, in alkalophiles the energy source is the transmembrane electrochemical gradient of Na^+[27] and swimming speeds are dependent upon extracellular Na^+.

Bacteriophages

Several phages have been described and a DNA phage has been studied in detail.[33] Predictably, phage multiplication is optimal at the optimal pH for bacterial growth (pH 9–11). Interestingly, the phages are stable in this pH range but unstable at pH 7.0 or less. What, if any, modifications to the phage coat are responsible for this are not known.

Cell walls

Alkalophiles appear to be structurally indistinguishable from neutrophiles. Thin sections of an alkalophilic *Bacillus* appear to be perfectly normal in the electron microscope, although the outer surface layer is complex, having some 'fuzzy' projections.[58] The constituents of the cell walls of alkalophilic bacilli have been compared with non-alkalophilic bacilli and the alkalophilic bacilli have characteristically three sorts of cell wall composition.[2]

1. large amounts of glucuronic acid and hexosamine
2. large amounts of glutamic acid with some aspartic and glucuronic acid in some strains
3. cell walls that contained similar amounts of phosphorus and neutral sugars as *B. subtilis* (a neutrophile) and were essentially identical.

Group 1 and 2 organisms contained less phosphorus and neutral sugars than *B. subtilis*. This suggested that the cell walls of some alkalophiles can have larger amounts of negatively charged materials. It was also found that growth of organisms at high pH increased the content of acidic compounds[2] and an alkalophilic *Bacillus* grown at pH 7.0 contained less aspartate, glutamic acid and glutamate than when grown at pH 10.0.[1,2]

Another difference between the groups was that Na^+ was not necessary for growth of group 3 organisms at pH 7 or 10.2. Groups 1 and 2 could not grow at pH 7.0 and Na^+ was essential for growth at higher pH values. Whether Na^+ acts to stabilize the increased amounts of negative charges in the cell walls is not known.

The peptidoglycan of ten alkalophilic bacilli was found to be very similar in composition to *B. subtilis* and was not affected by growth pH.[4] However, it has been shown that the teichuronic acid did form a greater proportion of the peptidoglycan in walls of bacteria grown at alkaline pH[1] and the molecular weight was apparently greater at alkaline pH (70 kDa) than at neutral pH (48 kDa). The cell walls are therefore composed of γ peptidoglycan, teichuronic acid and a polymer of glucuronate and glutamate. An unusual component of the teichuronic acid was found to be D-fucosamine.[3] The cell walls of alkalophilic and alkalotolerant actinomycetes have also been studied. Some contain *meso*-diaminopimelic acid whereas in general, streptomycetes have the LL form.[72]

In conclusion, the cell walls and flagella of alkalophiles and alkalotolerant organisms studied to date appear to be for the most part very similar to neutrophilic organisms. Some unusual components have been found but the most usual difference, which may not be obligatory but merely an aid to growth at alkaline pH, is an increase in the amounts of acidic materials.

Cell membranes

Turnover of phosphatidylglycerol (an important structural component of the membrane) is a general feature which occurs at a considerable rate in many bacteria[54] but in alkalophiles this turnover appears to be less extensive. The phospholipid content of alkalophilic bacilli was found to be very similar to that of *B. subtilis* but some unusual additional phospholipids were isolated in trace amounts. However, one, bis(monoacyglycero)phosphate is not usually found in microorganisms but composed up to 4% of the polar lipid of several alkalophilic bacilli.[78]

The membrane lipids of two obligate and 'facultative' bacilli have been compared in detail to those of *B. subtilis*.[8] The obligate alkalophiles contained high ratios of membrane lipid to protein and the lipid fraction had high amounts of neutral lipid. However, this was not thought to be a prerequisite for growth at high pH as the facultative strains were similar. The membrane diether core lipids of the halophilic alkalophiles, *Natronococcus* and *Natronobacterium*, have been shown to be unusual in that they contain C_{20}, C_{25} as well as the more usual archaebacterial C_{20}, C_{20} core lipid.[20]

Alkalophilic bacilli can also contain squalene and isoprenoids (β-carotene, phytoene) and have high concentrations of anionic phospholipids especially cardiolipin. Obligate alkalophiles contain high concentrations of branched fatty acids compared with *B. subtilis* and a high content of unsaturated fatty acids. Facultative alkalophiles contained almost no unsaturated fatty acid and a lower level of branched fatty acids than either *B. subtilis* or obligate alkalophiles.[8] In this study, polar lipids (glycolipids and phospholipids), which are common in other Gram-positive bacteria, were notably absent. Also, except for the phosphatidyl-ethanolamine, all of the phospholipid was anionic giving the membrane a high negative charge. Halophilic alkalophiles also contain carotenoids.[20]

Whether any of these changes in membrane composition are necessary for growth at alkaline pH is not certain but polar lipid compositions of phylogenetically diverse organisms show similar properties.[20] Some of these findings[81] suggest that the membranes have low melting points and hence are rather fluid. This may be linked to inhibition of the growth of alkalophiles at low pH due to problems of poor membrane integrity[59] and the apparent increase in K^+ permeability at acidic pH.[55] This may also account for the rarity of thermophilic eubacterial alkalophiles.

Changes in the membrane proteins that could be exposed to the pH_o have not been widely investigated but in one study it was shown that in an alkalotolerant *Bacillus*, grown at pH 7.5 and 10.2, there were quantitative changes in cell membrane proteins[20] and a greater negative charge on proteins was observed in cells grown at pH 10.

Respiration and oxidative phosphorylation

With some alkalophilic bacilli peak rates of oxygen consumption occur at pH 9–10[33] when amino acids are used as substrates, but carbohydrate stimulated respiration (glucose, glycerol) and endogenous respiration do not vary over the range pH 7–11. This perhaps reflects the maximal rates of transport of some amino acids often observed at alkaline pH (see later). With other strains of alkalophilic bacilli maximal respiratory activity is always observed around the pH optimal for growth (*c* pH 9.0).

The respiratory chains of alkalophilic bacilli are apparently normal, containing *a*, *b* and *c* type cytochromes and the rates of respiration are similar to non-alkalophilic bacteria.[59] At pH 9.0, NADH will cause a reduction of all cytochromes in membrane preparations but at pH 7.0 the *b* type cytochromes are not reduced. The NADH dehydrogenase from these alkalophiles has been purified and studied in detail and its activity is markedly stimulated by K^+ and Na^+ ions and to a lesser extent by phospholipids.[28] Quinones of alkalophiles have not been extensively studied but unsaturated and dehydrogenated menaquinones have been described in alkalophilic halophiles.[10] *Ectothiorhodospira* contains Q8 as the sole ubiquinone but menaquinones (MK8 and MK7) can also be present.[10,94] *B. firmus* RAB has been shown to contain menaquinone.[50]

A remarkable feature of alkalophilic bacilli is that they contain high levels of respiratory chain components, the concentrations of cytochromes being exceptionally high. This is responsible for the often observed intense pigmentation of these organisms. In *B. alcalophilus* and *B. firmus* there are many different cytochrome species as revealed by different mid-point potentials in difference spectra of their membranes — especially *b* type cytochromes.[50] Non-alkalophilic mutants of *B. alcalophilus* and *B. firmus* both have lower overall quantities and fewer distinct cytochrome species.

The respiratory proton pumping ability of *B. firmus* and its non-alkalophilic mutants have been studied.[66] At pH 9.0 the wild type had much higher H^+/O stoichiometries (between 9 and 13) than did the mutant at pH 7.0 (about 4) but the parent had a similar H^+/O ratio at pH 7.0. The significance of these observations is not clear. This may represent an adaptation that enables the organisms to meet unusual energetic costs of life at high pH[50] but this appears to be paradoxical as intracellular acidification, not high rates of proton extrusion, is required (see earlier). Furthermore, this cannot be advanced as an essential feature for all microorganisms as a facultative anaerobe that contains no cytochromes or quinones has been found which is alkalophilic (pH_{opt} 9.5–10, pH range 8–10) and appears to be able to maintain pH_o acidic with respect to pH_o.[57]

When the components of the Δp of alkalophilic bacilli were measured it was found that pH_i is more acidic than pH_o so that the ΔpH component was generally oriented in the opposite direction to that shown by neutrophiles and acidophiles. At pH_o 11, for instance, which permits good growth, the pH_i of *B. alcalophilus* is about 9.[13] The $\Delta\Psi$ component, as measured by the distribution of phosphonium ions, was found to be about 135 mV (inside negative), giving a net Δp of only 15 mV. This value is much lower than those observed in other organisms and according to chemiosmotic principles is clearly inadequate to drive ATP synthesis or active transport.

This observation has contributed to the continuing debate as to whether the energy coupling to ATP synthesis is strictly chemiosmotic and the bulk phase Δp is solely of energetic importance or that there is a localized flow of protons from the respiratory chain to the ATPase.[13] An alternative suggestion is that very high H^+/ATP stoichiometries occur at alkaline pHs. The explanation for these observations is quite unclear, although a distinct possibility is that current methods of measuring $\Delta\Psi$ give underestimates of its value;[6] the alternative is to accept that non-chemiosmotic mechanisms of energy coupling occur in these organisms. In other experiments[22] the ability of respiration and K^+ diffusion potentials to drive the synthesis of ATP in *B. firmus* were studied. At pH_o 7 with both $\Delta\Psi$ generation methods the ATP content increased but with respiration the increase was greater. In both cases ATP synthesis did not occur when the ATPase was chemically inhibited. In contrast, at pH 9, the K^+ diffusion potential was completely unable to drive ATP synthesis whereas malate driven respiration made more ATP than at pH 7, and in both cases the anticipated values of $\Delta\Psi$ could be measured. Furthermore, the K^+ diffusion potential was quite capable of driving AIB (α-aminoisobutyric acid) uptake at pH 9. These findings suggest that there is some (as yet undefined) direct link between respiratory derived protons and ATP synthesis; no other reasonable explanations have so far been advanced.

For the present, we must assume that oxidative phosphorylation in alkalophilic bacilli proceeds in a 'normal' Mitchellian fashion, protons, not Na^+ ions, are extruded by the respiratory chain and the bulk phase Δp drives ATP synthesis, but these as yet unexplained aspects must be borne in mind.

The Mg^{2+} dependent H^+ translocating ATP synthetase of alkalophilic bacteria appear to be perfectly normal in all respects so far examined[25] and increased stoichiometries of the ATPase at alkaline pH seem unlikely.

It could be suggested that this apparent paradox of having to pump protons out of the cytoplasm by respiration to drive ATP synthesis, and at the same time having a mechanism to bring extra protons into the cell to maintain pH_i more acidic would be costly in energetic terms[59,62] and alkalophiles would grow poorly. On the contrary, obligate alkalophiles (e.g. *B. alcalophilus* and *B. firmus*) have comparable molar growth yields on malate to non-alkalophiles (42 and 39 g dry wt mol^{-1}, respectively) at pH_o 10. Interestingly, the growth yield of their non-alkalophilic mutants, at their optimal growth pH 7.0, was no more than half that of the parent but this might be related to the pleotropic effects induced by these mutations.[59]

Active transport

The amino acid analogue α-aminoisobutyric acid (AIB) has been used extensively in studies of the active transport of amino acids as it is not incorporated into cellular material. Thus levels of cell-associated radioactive AIB should be due entirely to the properties of any AIB transporting active transport system(s). The transport of AIB into both whole cells and membrane vesicles of alkalophilic bacilli have been studied in detail using different metabolic substrates and inhibitors. These have shown that uptake of AIB is dependent upon an 'energized' membrane and that the rate and extent of uptake is greater at pH 9.0 than at pH 7.0.[59,62] This uptake is also dependent upon extracellular Na^+ ions and other cations, notably Li^+, could not generally substitute for Na^+ ions. Extracellular Na^+ ions were shown to increase the affinity of the uptake system for AIB transport[59] and compounds that dissipated $\Delta\Psi$ or the Na^+ gradient inhibited its uptake.

The uptake of other amino acids is also stimulated by sodium (glycine, L-alanine, L-serine, L-aspartate) but others appear not to be (proline, phenylalanine, methionine, valine, lysine). Glutamate, glucose and acetate are also co-transported with Na^+[33] as is efflux of Ca^{2+} ions which is driven by the trans-membrane Na^+ gradient via an antiport mechanism which exhibits optimal activity at alkaline pH.[78] Using Na^+ rather than protons as the coupling ion is not as unusual as first thought, and coupling to the Na^+ gradient is also observed in several marine bacteria and in *E. coli* for certain substrates (glutamate, melibiose, proline). Other transport systems, where studied, e.g. for lactose, are transported by some mechanism that employs ATP (directly or indirectly).[59]

An interesting observation was seen when the active transport characteristics of non-alkalophilic mutants of *B. alcalophilus* and *B. firmus* were examined. These mutants were shown to have lost Na^+/H antiport activity but also lost the ability for Na^+ gradients to drive the uptake of AIB and other Na^+ coupled solutes and the non-alkalophilic mutant of *B. alcalophilus* now exhibited co-transport of AIB with H^+ ions. A possible explanation[59] is that all Na^+-linked translocating porters share a common subunit that was lost by this mutation and that they then became H^+ linked. No other adequate explanations for these observations have yet been advanced and this hypothesis has not been developed further.

Protein synthesis and enzymes

Ribosomes from alkalophilic bacilli appear to be normal prokaryotic 70S ribosomes with normal thermal denaturation characteristics.[33] Indeed, there appears to be nothing exceptional in the whole protein synthesizing machinery as measured by the incorporation of radiolabelled amino acids except that in whole cells protein synthesis is more active at alkaline pH (7.6–10). However, cell free systems do show maximal activity at between pH 8 and 8.5, but this is only 0.5 pH units higher than that observed for a neutrophile, *B. subtilis*.[33] Of the few intracellular enzymes of alkalophiles so far examined, these appear to be very similar to corresponding enzymes of non-alkalophilic bacteria and have optimal activity around neutrality.[33] This is presumably due to the ability of alkalophiles to regulate pH_i towards neutrality. One enzyme, the β-lactamase of an alkalophilic *Bacillus*, has been studied in detail and the nucleotide sequence of the gene encoding the enzyme has been determined.[43] The gene has an open reading frame of 257 amino acids and a signal peptide of 30 amino acids of which 12 are hydrophobic and is very similar to other signal peptides that have been described.

Polyamines

Six different polyamines have been found in alkalophilic bacilli, three of them being cadaverine, spermine and spermidine.[33] High levels are observed in log-phase cells and ratios of spermine to spermidine have been shown to be higher in cells grown at pH 10 than at pH 7.0. The significance of these observations is not known.

ALKALOTOLERANT ORGANISMS

Although a great deal of effort has gone into understanding mechanisms of alkalophily, possible mechanisms of alkalotolerance have been largely ignored (except in *V. alginolyticus*). As such diverse types of organisms are alkalotolerant, it is probable that a wide range of cellular characteristics contribute to alkalotolerance. Whether any special mechanisms are involved is unclear but, like alkalophiles, in the alkalotolerant cyanobacterium *Synechococcus leopoliensis* Na^+ is involved in the resistance to alkaline pH. Cell division does not occur at any pH in the absence of Na^+ but cell activity and photosynthesis did continue at neutral pH values.[74] When cells were shifted to pH 9.6, only in the presence of Na^+ was pH_i maintained more acidic.

It could well be, with other organisms, that modification or special properties of the cell walls and membranes are required to allow the organisms to survive but not grow, or grow only slowly, at alkaline pH, and that there are no special mechanisms for holding pH_i at more acidic values. This is obviously an area worthy of further study.

Forced adaptation of bacteria to extremes of pH
There have been few reports of bacteria being able to extend their pH limits by applying selective pressure. Previous work[65] has described a strain of *B. cereus* which, by multiple transfers at increasing values of pH, could grow at pH 10.3. This characteristic became stable and was not lost on subsequent transfers. More recently[89] lactic acid bacteria have been shown to be capable of initiating growth at pH 10.0 by gradually increasing the starting pH values during enrichment runs. In these cases substantial amounts of acid was produced by the cultures to lower the pH but growth at the higher pH values was possible when the pH was subsequently raised. It is quite unclear what genetic, physiological or structural changes are occurring but this is obviously an area worthy of more research.

Furthermore, 174 strains of alkalophilic *Bacillus* (pH optimum 9.7) were shown to be capable of growth at pH 7.0 by successive transfer on media of decreasing pH values.[18] How this relates to naturally occurring facultative alkalophiles[21] also capable of growth between pH 7/7.5 and 10.5/12 is not clear.

Commercial aspects

A major limitation to the direct application of alkalophilic microorganisms to industrial processes is that there are only a few that at present are operated at alkaline pH. However, alkalophiles could be useful sources of materials for processes that do not operate at alkaline pH (e.g. antibiotics) or some processes could be adapted to use alkalophiles, although for the latter the benefits of using alkalophiles might have to be substantial if the cost of modifying the process is significant. Therefore, to date, the exploitation of alkalophiles has been confined to those processes that already operate under alkaline conditions.

The most extensively pursued aspect of the potential applications of alkalophiles has been the study of the many extracellular enzymes produced by alkalophilic bacilli which characteristically exhibit optimal activity and stability at alkaline pH. These have been described in detail[33,88] and will be only summarized here.

ENZYMES OF ALKALOPHILIC ORGANISMS

Proteases
The most well-known and used proteases are the subtilisins produced by *B. subtilis* and other closely related neutrophilic, and even thermophilic,[69] bacilli. These enzymes are stable over a pH range of 5–10 and have optimal activity at alkaline pH. The subtilisins have been divided into two distinct groups based on their amino acid composition but they do show over 60% homology in their primary amino acid sequences.[82] These enzymes are active only at low temperatures and are inactivated by high alkalinity and temperatures in the absence of Ca^{2+} ions.

Many alkalophilic microorganisms can also produce alkaline proteases in large amounts. When studied in detail many of these proteases were found to be similar in several respects to the subtilisins, e.g. having similar sensitivities to different inhibitors and a stimulation and stabilization of activity by Ca^{2+} ions at elevated

(60 °C) temperatures, but the alkalophilic proteases did have higher isoelectric points and specific activities.[33] For example, an alkaline protease from an alkalophilic *Bacillus* sp. has been studied in detail and shown to have a pH optimum of 10–11 for the digestion of milk casein.[93] The enzyme also showed a unique specificity for the oxidized form of the β chain of insulin. The activity of this enzyme was completely lost at 50 °C but activity was completely retained at 45 °C by the presence of Ca^{2+}.

In general, proteases from alkalophilic bacteria have higher pH optima than subtilisins and fairly broad pH activity spectra (e.g. pH 8–12) and higher temperature stability and optima (60–65°C). However, other novel alkaline heat stable proteases have been isolated which do not require Ca^{2+} ions for activity,[12] which suggests that these are quite different enzymes.

Elastin is a flexible, cross-linked protein present in some mammalian tissues, e.g. lungs, arteries. This protein is insoluble and unusually stable, being very resistant to enzymatic degradation. However, several elastolytic enzymes from micro-organisms have been isolated including an elastase from an alkalophilic *Bacillus*[92] where it might be expected that elastolysis might be naturally enhanced by the high pH. This enzyme showed a marked preference for elastin over other proteins but the enzyme would also hydrolyse keratin and collagen. Like many alkaline proteases, the enzyme required Ca^{2+} for stability but interestingly was inhibited by Na^{+} ions. The optimal pH for activity was pH 11.7 but the enzyme showed a fairly broad activity from pH 8.5–12.7 and had optimum activity at 60 °C. The amino acid sequence showed some homology with subtilisins and was sensitive to general protease inhibitors.

Applications of proteases

These features of stability at elevated temperature and pH of alkalophilic proteases led to the development of one of the most notable industrial applications of the products of alkalophiles.[33] Several alkalophilic proteases were shown to be very stable, albeit to differing extents, in the presence of a variety of detergents. The enzymes were tested in detail as to their ability to enhance the detergent action in the washing of soiled clothes. They proved to be extremely effective and, after some initial fears of possible allergic reactions, this has led to their commercial use and production world wide. Two such commercially available enzymes are Alcalase^{R} and Esperase^{R}, both produced by bacilli.[88] As well as the aforementioned properties, the enzymes were particularly suitable because they could be produced from bacteria with high yields and the enzymes were soluble and could be stored for long periods of time.

Another industrial process which has received attention is the enzyme-assisted de-hairing of animal hide. Traditionally, this process is carried out by treating animal hides with a saturated solution of lime and sodium sulphide. Besides being expensive and particularly unpleasant to do, a strongly polluting effluent is produced. Enzyme-assisted de-hairing was obviously possible if proteolytic enzymes could be found that were stable and active under the alkaline conditions (pH 12) used. Early attempts using a variety of enzymes were largely unsuccessful but proteases from alkalophilic bacteria have been shown to be highly effective in assisting the hair removal process.[33,88] Several alkaline proteases from alkalophilic actinomycetes have also been isolated and studied.[73] Some of these have been shown to be particularly active against keratinous proteins such as hair, feathers, wool, etc. at alkaline pH and could have commercial applications. One of these proteases is also active against peptide linkages in oxidized insulin and the specificity of the enzyme is very different from other alkaline proteases.[88]

Alkaline proteases, i.e. subtilisins, are also widely used in the food industry but alkaline proteases from alkalophilic organisms do not appear to have been applied, perhaps due to legal restrictions on the sources and types of enzymes that can be used due to possible toxic effects.

Cyclodextrin glycotransferase (CGTase)

The cyclodextrins (CDs) are a group of homologous oligosaccharides produced from starch by the action of CGTase. Cyclodextrins are cyclic molecules composed of six (α-CD), seven (β-CD) or eight (γ-CD) α-D-glucopyranose units linked by α-1,4 linkages. They are unusual carbohydrates as they do not have reducing or non-reducing end groups, which makes them fairly resistant to acid or alkali and the common starch degrading enzymes (amylases). They will crystallize easily from aqueous or alcoholic solutions and can form inorganic complexes with neutral salts, halogens and bases.

These properties give CDs many wide potential applications including the stabilization of volatile materials, protection of sensitive components against oxidation and UV degradation during storage or processing of products or raw materials, modification of physical and chemical properties of substances, emulsification or solidification of fats, oils and fatty acids. In particular, CDs could be used in foods as emulsifiers, foaming agents or flavour stabilizers; in cosmetics as colour or fragrance stabilizers or maskers; in pharmaceuticals for increasing the solubility or stability of antibiotics, hormones or vitamins; in pesticides, plastics and many other areas.[33]

CGTases have been found in neutrophilic bacteria mostly as extracellular enzymes. They differ mainly in the ratios of the amounts of the different types of CD they will produce. The CGTases of *B. macerans* have been used widely for the production of CDs but these enzymes, like subtilisins, suffer from a lack of thermal stability, again an important drawback for commercial exploitation. Thermostable CGTases have been isolated from alkalophilic bacteria growing at pH 9.5–10.3. The enzymes have broad pH activity (5–9) with optimal activity at pH 8–9. These crude CGTases have been shown to convert potato starch to CDs with a yield of about 70% at pH 4.7 and 75% at pH 8.5. This activity is in fact due to a mixture of enzymes — acidic, neutral and alkaline CGTases — all of which produce the different types of CD at different rates. The acid, neutral and alkaline CGTases from an alkalophilic *Bacillus* sp. have been immobilized on synthetic resin with no loss of activity for two weeks of continuous operation although activity did decline to about half after 25 days.[40] No change in the thermal stability was observed (55 °C optimum) but the acidic CGTase was inactivated on immobilization shifting the optimal enzyme activity to around pH 9. The resin was contained in a glass column and a substrate of 4% w/v starch (in Tris Buffer pH 8.5) was pumped through at various flow rates. The immobilized CGTase converted about 63% of the soluble starch to CD. Thus large-scale production of CD with high yields from starch can now be achieved and some of the potential applications detailed above are now in use.[88]

The nucleotide sequence of a β-CGTase of an alkalophilic *Bacillus* has now been determined after cloning the gene into a plasmid that was stably maintained in *E. coli*.[48] More than 70% of the enzyme activity was expressed in the culture medium and the enzyme activity was similar to that in the donor strain; the major product from the hydrolysis of starch was β-CD. An interesting aspect was that the enzyme showed a high degree of homology at the N-terminal end with α-amylases ('normal' starch degrading enzymes). However, the C terminal end of the β-CGTase was completely different from α-amylases and this region of the β-CGTase contained an extra 200–250 amino acid sequence in addition to the area having the amylase activity. This suggested that the β-CGTase may consist of two active domains. The N terminal end may cleave the α-1,4 glucosidic bond in starch (like α-amylases) and the other region catalyses the reconstruction of the linkage to produce the cyclic oligosaccharide.

Amylases

Several α-amylases have been isolated from alkalophilic bacilli and they can hydrolyse 30–70% of starch to yield glucose and maltose and maltotriose to a lesser extent.[88] The enzymes have typical pH and temperature optima of 10.5 and 55 °C and they are remarkably resistant to EDTA. They have also been classified into four groups depending on the pH activity profiles and other features, suggesting that there are indeed several distinct types of enzyme. These enzymes are becoming increasingly important as detergent additives and in the food industry for removing residues from starch-containing foods.[88] One alkalophilic amylase has been found to have the unusual ability to hydrolyse β-cyclodextrin to give maltotriose and glucose.[88] Three unusual alkaline amylases have been found in an alkalophilic *Bacillus* as they produce maltohexose as the main product from starch.[24] The enzymes all had similar activities and were stable in the pH range 7–12, and had optimal temperature and pH values of 60 °C and 10.5 respectively. Amylases with alkaline pH optima have also been found in alkalophilic actinomycetes.[73]

Cellulases

Renewable cellulose biomass is a potentially important resource for the production of useful fuels and chemicals and there has been considerable interest in cellulases of microbial origin. Many cellulases are produced by mesophilic fungi with optimal activity at pH 4–5 and 40–50 °C. Alkaline and partially alkaline cellulases have been discovered but they have not yet been successfully applied for industrial use.[33]

An alkalophilic *Aeromonas* sp. produces cellulase with optimal activity at pH 8.0. The cellulase is a mixture of at least four enzymes (but see later) and acts on cellulose, filter paper and other materials and its potential use in sewage processing has been investigated.[88] An alkalophilic *Bacillus* sp. was also found to be a good producer of alkaline cellulase and this also was a mixture of at least four enzymes.[33] The enzymes had a broad pH activity (4.5–11) but optimal activity was between pH 6–7; it was, however, stable up to 50 °C. Two other cellulases with even higher temperature stability have been isolated from alkalophilic bacilli.[14] The enzymes again had broad pH activity (5–10) but these enzymes both had optimal activity at pH 10.0. One of the enzymes was stable up to 60 °C but the other was stable up to 80 °C.[34] Cellulases have also been demonstrated in thermophilic anaerobes isolated from an alkaline hot (75–85 °C) spring.[91] These organisms had a pH optimal for growth at about pH 7.5–8.0 and an orange pigmentation. These *Clostridium* sp. were shown to be effective in decomposing crystalline cellulose, rice straw, newspaper and bagasse, producing ethanol and acetate as the major end-products.

Two genes coding for the cellulases of an alkalophilic *Bacillus* have been sequenced[16] and shown to code for proteins containing 409 and 488 amino acids. The two proteins show partial homology but one of them contains a direct repeat sequence of 60 amino acids which was shown to be not necessary for cellulase activity. It is thought that the DNA codes for multiple cellulase genes that may have arisen by gene duplication.

Xylanases

Xylan is a carbohydrate component of plants composed mainly of β-1,4-D-xylose in land plants and β-1,3-D-xylose in marine plants. Like cellulases, xylanases have attracted considerable attention as a means of converting biomass into valuable resources, as xylan is abundant in agricultural and other wastes (e.g. straw).

Not surprisingly, xylanases are widely distributed in bacteria and fungi and some have been isolated from alkalophilic bacteria. Maximal enzyme production has been shown to be dependent on carbonate in the medium (reason unknown) and the enzymes have optimal activity between pH 6 and 8. The maximum xylan hydrolysis is only about 40% at either pH 6 or 9.[33] However, xylanases from alkalophiles do retain higher activity at alkaline pH than other bacterial xylanases.

One strain of an alkalophilic *Bacillus* has been shown to produce two types of xylanase. Xylanase N is most active in the pH range 6–7 whereas xylanase A has much broader activity (pH 6–10) and is still active at pH 12.0.[31] Both enzymes convert xylan into xylobiose and other oligosaccharides but with efficiencies of only 25%, but they cannot hydrolyse xylobiose or xylotriose. Thermostable xylanases have also been isolated from four strains of moderately thermophilic (45–50 °C), alkalophilic (optimum pH 9–10) bacilli from soil.[79] These organisms produced xylanases only when xylan or xylose was present in the medium. The optimal conditions for these enzymes were a neutral pH (6–7) and temperatures between 65 and 70 °C. With these enzymes the major products from xylan were xylose and xylobiose.

There have been no practical applications to date of alkalophilic xylanases. Possibilities do exist, e.g. alkalophilic bacilli are capable of utilizing rayon waste[33] due to xylanase activity giving rise to xylose, xylobiose and oligosaccharides as the principal products. Perhaps the thermostability exhibited by the xylanases from some alkalophilic bacteria[79] might have more potential.

Pectinases
Pectic polysaccharides are common in higher plants and pectinases from some microorganisms are widely used in fruit and vegetable processing. Pectinases have been isolated from alkalophilic bacilli, the optimum of the enzyme pH generally being about 10. There appears to be little practical use for these enzymes in fruit and vegetable processing as these are usually acidic. However, the pectic lyase (pH optimum 9.5) produced by an alkalophilic *Bacillus* has been used in improving the production of a type of Japanese paper.[97] A new retting process was developed exploiting the strong macerating ability of the enzyme on the raw material and this produced a high quality, non-woody paper which was stronger than the paper produced by the conventional soda and ash cooking method.

Pectinolytic enzymes are usually considered to be hydrolases. However, pectin attacking enzymes have been discovered which catalyse the eliminative degradation of pectin. Such an enzyme, polygalacturonate lyase, has been isolated from an alkalophilic *Bacillus*.[44] Highest enzyme production was shown to occur in a complex nitrogen medium with an initial pH of 9.7. The pH optimum of this extracellular enzyme in buffer was 10.0 and activity was activated by Ca^{2+} ions.

Other enzymes
Several other enzymes from alkalophilic or alkalotolerant organisms have been studied but have not as yet found a commercial application. They are, however, of interest due to their alkali resistance when they are extracellular enzymes.

Pullulan is a polysaccharide produced by yeasts. An alkaline pullulanase has been isolated and studied in detail.[33] The enzyme gave complete hydrolysis of amylopectin or glycogen when used with β-amylase, whereas the use of β-amylase alone resulted in only partial degradation. β-1,3-Glucan is a common polysaccharide component of cell walls in microorganisms and higher plants. Many microorganisms contain β-1,3-glucanases that have strong lytic activity on the cell walls and spore coats of fungi.[33] Most have pH optima <7 but thermostable enzymes with activity at pH values in excess of 8 have been isolated from alkalophilic bacteria.

β-Lactamases are important enzymes in antibiotic resistance mechanisms and chemotherapy as they can inactivate penicillins or cephalosporins by hydrolysis. Not surprisingly, an alkalophilic *Bacillus* has been shown to produce an inducible extracellular β-lactamase at alkaline (9.0) pH. The optimum pH of the enzyme was between pH 6 and 7. The nucleotide sequence of the β-lactamase from an alkalophile has been determined.[43] The gene contains an open reading frame coding for 257 amino acids which represents the precursor β-lactamase protein. This precursor contains a thirty amino acid peptide which is very similar to other signal peptides and includes twelve hydrophobic amino acids.

Both extracellular and intracellular lipases are found in many microorganisms. Extracellular lipases have been found in several alkalophilic bacteria with optimal activities at pHs between 9.5 and 10. Alkaline lipases (pH optimum 8–9) have also been found in fungi.[33] Lipases could have applications in detergent formulations.

A *Penicillium*-like alkalotolerant organism has been shown to produce polyamine oxidase extracellularly (it is usually an intracellular enzyme) which catalyses the oxidation of spermine and spermidine to putrescene and other chemicals,[33] but the optimum pH of this enzyme was pH 4.0. A uricase with a pH optimum of 9.0 has been isolated in the culture fluid of an alkalophilic streptomycete and an alkaline (pH 9.0 optimum) alginate lyase, DNAase and RNAase have been isolated from alkalophilic bacilli.[33] Catalases are also found in alkalophilic bacilli but an unusual extracellular catalase has been found which has an optimum pH of 10.00 and is fairly heat stable; this could have several industrial applications. α-Glucosidase hydrolyses α-1,4 and/or α-1,6 glucoside linkages in short-chain saccharides that arise from the action of other enzymes on starch. The extracellular α-glucosidase activity of an alkalophile has been shown to be due to the action of two enzymes, an α-glucosidase and a maltase. Both enzymes had pH optima at 7.0, unlike many other extracellular enzymes of alkalophiles.[45]

Immobilized enzymes and cells

There have been few reports of the effects of immobilizing enzymes from alkalophilic bacteria but where done it has produced interesting and useful effects (see section on CGTases).

Another enzyme from an alkalophilic *Bacillus* that has been immobilized is dihydropyrimidinase. One strain was found to have high activity of this enzyme which was immobilized by entrapment in polyacrylamide gel.[96] Yields of up to 80% were observed when using various substituted hydantoins as substrates. Spontaneous racemization of hydantoins occurs under alkaline conditions and thus enzymes from alkalophiles could be very useful. This enzyme catalyses the hydrolytic ring opening reaction of dihydropyrimidines to produce N-carbamyl β-amino acids and has been used in an enzymatic process for the production of amino acids (D configuration) that are important components of semisynthetic penicillins and cephalosporins. Recently[71] the production of glutamate by a *Synechococcus* sp. immobilized in alginate spheres has been studied. Glutamate was 83% of the total amino acids produced and more than 95% of the glutamate produced was secreted into the medium at a constant rate for 7 days.

OTHER COMMERCIAL ASPECTS

The first traditional and industrial application of alkalophiles was in the 1960s. The indigo reduction process was improved by the addition of a culture of an alkalophilic *Bacillus* giving both a faster process and a better product.[33]

Sources of novel antibiotics

There is a continuing need to discover new antibiotics with chemotherapeutic properties against both bacterial and fungal infections. It is therefore not surprising that alkalophilic microorganisms should have been examined as potential sources for novel antibiotics, especially as alkalophilic and alkalotolerant streptomyces are known. Indeed, many different sorts of antibiotics with different activities and spectra of activities have been shown to be produced by alkalophiles. Some are produced only under alkaline conditions, but as far as is known none are in commercial use due to toxicity problems. However, as this area is so important commercially, antibiotics from alkalophiles could well be under development and have not been reported.

Actinomycetes generally grow best in neutral or slightly alkaline media but several truly alkalophilic actinomycetes have been isolated that produce β-lactam type antibiotics and nocardicins as well as several new antibiotics.[72] One alkalophilic streptomycete strain isolated from soil was shown to produce antimycin A in significant amounts[85] under alkaline (pH 10.3) conditions which was active against *Candida albicans*.

In one large study, 3000 strains of alkalophilic and alkalotolerant microorganisms, mainly streptomycetes and fungi, were screened as to their ability to produce antibiotics.[86] Several hundred were shown to have significant antifungal or antibacterial properties (or both) which were produced under alkaline conditions. Strong antifungal activity was found in fifteen streptomycetes and was shown to be antimycin A. One fungal strain produced a potent antibiotic against Gram-positive cocci only (helvolic acid). Another fungal strain produced antibiotics with broad and potent antimicrobial properties against yeast, bacteria and fungi. These appeared to be a homologous group of peptide antibiotics with novel structures but many antibiotics have not been characterized as yet.[86]

Detrimental aspects

Commercial problems due to alkalophiles have not been reported often. Recently, the production of malodours in buildings where casein was used as an additive to improve the fluidity of the concrete (pH 12–12.5) used in construction has been ascribed to alkalotolerant clostridia.[38] These organisms could be isolated from the concrete and were shown to produce various volatile organic acids and amines when grown on casein.

Genetic aspects

In truth, knowledge of the genetics and molecular biology of alkalophiles is almost non-existent. Apart from the isolation of a few mutants (for elucidating pH_i regulation mechanisms), nothing is known of the genetics of alkalophilic bacilli or other alkalophiles. No detailed genetic maps have been constructed as for *E. coli* and *Salmonella typhimurium*. Even the occurrence of plasmids and the general organization of the genetic material and the location of particular genes is not well characterized, nor has the existence of genetic transfer systems been described, although it is probably reasonable to assume that such systems do exist. Thus, there are exciting opportunities for future research in these aspects and they will undoubtedly be pursued in the future.

It is perhaps inevitable that the techniques of molecular biology, developed in other organisms, have been applied to alkalophiles (mainly bacilli) and a significant number of genes from alkalophiles have been cloned into neutrophilic hosts. This could have advantages such as increased levels of expression of the required enzyme in the new host, alterations in the properties of the enzyme and excretion of normally intracellular enzymes into the medium. It could also offer possibilities for protein engineering to subtly or substantially alter the properties of the enzyme. Obviously, target enzymes with direct commercial potential have been investigated initially (Table 3.2).

Table 3.2 Cloning and expression of alkalophilic genes in non-alkalophilic hosts

Target gene	*Alkalophilic donor*	*Plasmid vector*	*Restriction endonuclease*	*Host organism*	*Comments*	*Reference*
Cellulase	*Bacillus* sp.	pBR322	*Hind* III or *Eco* R1	*E. coli*	Cellulase activity expressed in periplasm. Same pH range as donor	42, 84
Cellulase	*Bacillus* sp.	pEAP37	*Hind* III or *Eco* R1	*E. coli*	Most of cellulase activity in culture	42
Xylanase	*Aeromonas* sp.	pBR322 pEAP37	*Hind* III or *Eco* R1	*E. coli*	Xylanase synthesis constitutive. Most xylanase activity found in the periplasm when using pBR322. Use of pEAP37 led to xylanase being in the medium	42, 64
Xylanase	*Bacillus* sp.	pBR322 pCX311[a]		*E. coli*	Most of xylanase produced in culture broth. Xylanase had a broad pH activity	31
Xylanase	*Bacillus* sp.	pBR322	*Hind* III	*E. coli*	Xylanase A cloned and was produced extracellularly at higher activity than donor	

Penicillinase	*Bacillus* sp.	pBR322	*Hind* III or *Eco* R1	*E. coli*	Penicillinase activity mainly in supernatant	51
β-isopropylmalate dehydrogenase	*Bacillus* sp.	pBR322	*Hind* III	*E. coli*	Activity low due to lack of promoter	41
β-cyclodextrin synthetase	*Bacillus* sp.	*E. coli* Φ λD69 *B. subtilis* Φ pUB110	*Sau* 3Al *Bam* H1	*E. coli* *B. subtilis*	High enzyme activity in supernatants. Major product of starch hydrolysis was β-cyclodextrin	49

[a]pBR322 containing a 4.6 kb *Hind* III fragment of a *Bacillus* sp.

CLONING OF GENES FROM ALKALOPHILES

In other microorganisms cellulase is usually a mixture of different enzymes. However, a cellulase gene from an alkalophilic *Bacillus* has been cloned into *E. coli* and appears to be a single enzyme having significant cellulase activity.[84] The activity expressed had the same pH profile as the donor strain and in this case the cellulase activity was strongest in the periplasm of the host. This was also observed when the xylanase from an alkalophilic *Aeromonas* was cloned into *E. coli* using pBR322 as the plasmid vector.[42] When a different plasmid vector (pEAP37) was used, most of the cellulase and xylanase activity appeared in the culture medium.[42]

Like cellulase, xylanase activity is usually a multi-enzyme system in microorganisms. However, a xylanase gene has been cloned from an alkalophilic *Bacillus* and an *Aeromonas* sp. into *E. coli* and is thought to be a single gene.[29,31,64] Both cloned enzymes had the same pH activity profile as in the donor organism and most of the enzyme activity appeared in the medium. The cloned xylanase from *Aeromonas* was constitutively expressed and its activity was eighty times higher than in the donor organism.

The xylanase activity of an alkalophilic *Bacillus* was studied in more detail[79] and the organism was shown to produce two xylanases, N and A, with pH optima of 7.0 and 6–10 respectively. Both enzymes are produced extracellularly but the potentially more useful enzyme A (wider pH optimum) consisted only of about 15% of the total enzyme in the medium. When the xylanase A enzyme was cloned into *E. coli* the enzyme was also produced extracellularly and had the same wide pH activity profile but the xylanase activity was much higher than in the donor organism. Xylanase production in *E. coli* was inhibited by glucose and the enzyme was active only in the presence of monovalent cations.[30] Furthermore, the β-lactamase of the *E. coli* strain carrying the cloned plasmid was now found in the extracellular medium (normally a periplasmic enzyme) suggesting that the plasmid had induced some permeability changes in the host organism. These changes must have been quite subtle, however, as the β-galactosidase (cell associated) and most of the alkaline phosphatase (normally located in the periplasm) activity of the *E. coli* were still mainly located in the cell and periplasm respectively.

The penicillinase from alkalophilic *Bacillus* sp. has been cloned into *E. coli*[51] and enzyme activity was mainly located in the medium (as in the donor strain) but, interestingly, alkaline phosphatase activity (normally periplasmic) was now found mostly in the medium. Another enzyme from an alkalophilic *Bacillus*, β-isopropylmalate dehydrogenase, has been cloned into *E. coli* but the activity of the cloned gene was low in the host organism, possibly due to the lack of a promoter site on the cloned fragment.[41] The β-cyclodextrin synthetase gene from an alkalophilic *Bacillus* has also been cloned into *E. coli* and *B. subtilis* (a neutrophile) using chimeric phages.[49] In this case, the cloned genes were shown to be stably maintained on plasmids in the hosts. High levels (95% in *B. subtilis*, 70% in *E. coli*) of enzyme activity was found in the media of the host strains. The major product of hydrolysis of starch was β-cyclodextrin; α-cyclodextrin and γ-cyclodextrin were also detected, i.e. the properties of the cloned enzyme were very similar to that of the parent.

The production of these enzymes in the culture medium is interesting. In the donor organisms these enzymes are naturally extracellular or periplasmically located enzymes, so it is perhaps not surprising that the enzymes were exported through the cytoplasmic membrane as they presumably contained signal peptides for this purpose. However, that they should be released past the outer membrane of the host *E. coli* was unusual. With the cloned penicillinase, only periplasmic proteins were released, not cell enzymes like β-galactosidase. This suggested that a DNA fragment of the alkalophilic organism has also possibly been cloned with the target genes, and that this fragment was apparently making the outer membrane of *E. coli* permeable, allowing periplasmic proteins such as penicillinase and alkaline phosphatase to be excreted into the medium.

One of the plasmids used for the cloning of the penicillinase gene was pMB9 which contained the *kil* gene, which is responsible for the release of colicin E1 by cells carrying colicin plasmids (pMB9 was derived from such a plasmid). However, the *kil* gene cannot be expressed in pMB9 because both the gene for colicin E1 production and its promoter have been deleted from the plasmid and both are required for *kil* gene expression.[51] It was shown that the *kil* gene was activated in the cloned plasmid by read through from a promoter in the inserted fragment and that this caused an increase in the permeability of the outer membrane of the *E. coli* host. The exact mechanism of this permeability change is not certain, but the *kil* gene product exhibits partial homology with the lipoprotein component of the outer membrane which suggests it could replace native lipoprotein and thereby affect the permeability of the outer membrane.

A further set of experiments were carried out on a cellulase enzyme from an alkalophilic *Bacillus* by cloning truncated genes into *E. coli*.[15] The complete cellulase of the *Bacillus* was shown to be an enzyme consisting of 770 amino acids with a molecular weight estimated to be of 99 kD. Interestingly, it was found that one of the truncated gene products, a protein of only 46 kD, had cellulase activity and a pH optimum very similar to that of the original enzyme. Furthermore, deleting thirty-two amino acids from the protein resulted in total loss of enzyme activity, identifying those as being important in cellulase activity, perhaps, but not necessarily, being near or at the active site of the enzyme.

A particularly relevant aspect, which requires further study, is that, like many extracellular enzymes, CGTase begins to be synthesized only at the end of the logarithmic phase of growth and fully produced only in the stationary phase (e.g. after 48–72 h growth). Knowledge of the control of the expression of these genes is obviously important in optimizing their production and it has been shown that this could be linked to a gene developmentally regulating spore formation, but the presence of glucose also had an effect.[32]

One other molecular biological aspect of alkalophiles is worthy of note. The DNA from an alkalophile has been shown to be capable of transforming *B. subtilis* (a neutrophile) to alkalophily.[90] This was shown by preparing purified DNA from strains of an alkalophile and using these to transform *B. subtilis*. Unfortunately, how many genes were involved in causing this change could not be demonstrated. This would require a more detailed genetic study using different recombinant methods which are as yet not directly applicable to alkalophilic bacilli, let alone other alkalophilic microorganisms. Other genes from alkalophiles may well have been cloned but not reported for commercial reasons; undoubtedly more will be in the future.

The development of new vector systems for alkalophilic bacilli is certainly possible. Plasmids have now been identified in these organisms but only two strains out of the 200 screened contained them,[32] but they do contain restriction sites, as one would expect, thus the potential for using them as vectors is certain.

Future prospects

The crystal ball is as yet a little hazy. There seems little doubt that new alkalophilic microorganisms will continue to be discovered and that these or already known organisms will have useful products or properties that can be exploited practically. Where the next great development will be is unclear but it may be that alkalophiles or their products will be starting points for new directions. For instance, the cyclodextrins could have most interesting properties as 'artificial enzymes'.[7] Being soluble in water but having a hydrophilic interior, hydrophobic molecules with the right shape can bind two molecules so that they can react. Careful design of the ring or the addition of active side chains can greatly speed up the catalytic activity of these molecules; the potential applications of this are enormous.

Molecular biology must have an important future and the cloning of enzymes from alkalophiles could be important in several respects. The amino acid composition and sequences of the active sites of some enzymes are known.[33] The examination of changes in the structure and activity of these enzymes by subtle alterations in the amino acid sequences using such techniques as site-directed mutagenesis could be beneficial in producing enzymes with enhanced pH or temperature stabilities or activities. It should also give fundamental insights into the mechanisms of enzyme action. Indeed, site-directed mutagenesis experiments have produced subtilisins with enhanced stability.[82] Whether there is a direct link between thermal and alkali stability is not known. These studies suggest that there is, and alkali and temperature stability can be, but may not necessarily be, conferred together.[82] These approaches, together with cloning of alkalophilic genes into other organisms and other aspects of manipulation of enzyme activity, e.g. by immobilization, will probably see the most activity in the coming years.

The physiological basis of alkalophily in the bacilli is now fairly well understood. It will be most interesting to see what common features and differences emerge when other alkalophiles are studied in as much detail.

Conclusions

We have seen that alkalophilic and alkalotolerant microorganisms are a diverse collection of unrelated organisms. It is tempting to speculate that the ability to exploit alkaline environments is probably not an early evolutionary trait preserved throughout a few descendent lines but that this ability has arisen many times throughout evolution and has been preserved when and where advantageous.

We really know little about the alkalophiles. The two aspects that have received most attention in recent years — the mechanisms of membrane transport and pH_i regulation, and the study and cloning of the extracellular enzymes of bacilli — have not only given insights into fundamental aspects of microbial life but have also produced exploitable commercial opportunities. It can only be concluded that the further study of alkalophiles will continue the development of these and other aspects.

Acknowledgements

The author would like to thank Ian Booth (Aberdeen) and Roy Patchett (Reading) for useful comments on the manuscript, and Hilary Woodford and Clive Edwards for arranging cluttered scribblings into, I hope, a readable chapter.

References

1. Aono, R. (1985) *Journal of General Microbiology*, **131**, 105–11.
2. Aono, R. and Horikoshi, K. (1983) *Journal of General Microbiology*, **129**, 1083–7.
3. Aono, R. and Uramoto, M. (1986) *Biochemical Journal*, **233**, 291–4.
4. Aono, R., Horikosho, K. and Goto, S. (1984) *Journal of Bacteriology*, **157**, 688–9.
5. Booth, I.R. (1985) *Microbiological Reviews*, **49**, 359–78.
6. Booth, I.R. (1988), in R. Whittenbury, G.W. Gould, J.G. Banks and R.G. Board (eds) *Homeostatic Mechanisms in Microorganisms*, Bath University Press, pp. 1–12.
7. Breslow, R. (1988) *New Scientist*, No. 1621, p. 44–7.
8. Clejan, S., Krulwich, T.A., Mondrus, K.R. and Seto-Young, D. (1986) *Journal of Bacteriology*, **168**, 334–40.
9. Collins, M.D., Lund, B.M., Farrow, J.A.E. and Schleifer, K.H. (1983) *Journal of General Microbiology*, **129**, 2037–42.
10. Collins, M.D., Ross, H.N.M., Tindall, B.J. and Grant, W.D. (1981) *Journal of Applied Bacteriology*, **50**, 559–65.
11. Dibrov, P.A., Kostyrko, V.A., Lazarova, R.L., Skulachev, V.P. and Smirn-Ova, I.A. (1986) *Biochimica et Biophysica Acta*, **850**, 459–67.
12. Durham, D.R., Stewart, D.B. and Stellwag, E.J. (1987) *Journal of Bacteriology*, **169**, 2762–8.
13. Ferguson, S.J. (1985) *Biochimica et Biophysica Acta*, **811**, 47–95.
14. Fukumori, F., Kudo, T. and Horikoshi, K. (1985) *Journal of General Microbiology*, **131**, 3339–45.
15. Fukumori, F., Kudo, T. and Horikoshi, K. (1987) *FEMS Microbiology Letters*, **40**, 311–14.
16. Fukumori, F., Sashihara, N., Kudo, T. and Horikoshi, K. (1986) *Journal of Bacteriology*, **168**, 479–85.
17. Gee, J.M., Lund, B.M., Metcalf, G. and Peel, J.L. (1980) *Journal of General Microbiology*, **117**, 9–17.
18. Gordon, R.E. and Hyde, J.L. (1982) *Journal of General Microbiology*, **128**, 1109–16.
19. Grant, W.B. and Tindall, B.J. (1980), in G.W. Gould and J.E.L. Corry (eds) *Microbial Growth and Survival in Extremes of Environments*, London and New York, Academic Press, pp. 27–36.
20. Grant, W.D. and Tindall, B.J. (1986), in R.A. Herbert and G.A. Codd (eds) *Microbes in Extreme Environments*, London, Academic Press, pp. 25–54.

21. Guffanti, A.A., Finkelthal, O., Hicks, D.B., Falk, L., Sidhu, A., Garro, A. and Krulwich, T.A. (1986) *Journal of Bacteriology*, **167**, 766–73.
22. Guffanti, A.A., Fuchs, R.T., Schneier, M., Chiu, E. and Krulwich, T.A. (1984) *Journal of Biological Chemistry*, **259**, 2971–5.
23. Harold, F.M. and van Brunt, J. (1977) *Science*, **197**, 372–3.
24. Hayashi, T., Akiba, T. and Horikoshi, K. (1988) *Applied Microbiology and Biotechnology*, **28**, 281–5.
25. Hicks, D.B. and Krulwich, T.A. (1986) *Journal of Biological Chemistry*, **261**, 12896–902.
26. Higgins, C.F. and Booth, I.R. (1988), in R. Whittenbury, G.W. Gould, J.G. Banks and R.G. Board (eds) *Homeostatic Mechanisms in Microorganisms*, Bath University Press, pp. 29–40.
27. Hirota, N. and Imae, Y. (1983) *Journal of Biological Chemistry*, **258**, 10577–81.
28. Hisae, N., Aizawa, K., Koyama, N., Sekiguchi, T. and Hosoh, Y. (1983) *Biochimica et Biophysica Acta*, **743**, 232–8.
29. Honda, H., Kudo, T. and Horikoshi, K. (1985) *Journal of Bacteriology*, **161**, 784–5.
30. Honda, H., Kudo, T. and Horikoshi, K. (1986) *Systematic and Applied Microbiology*, **8**, 152–7.
31. Honda, H., Kudo, T. and Horikoshi, K. (1986) *Journal of Fermentation Technology*, **64**, 373–7.
32. Horikoshi, K. (1986), in R.A. Herbert and G.A. Codd (eds) *Microbes in Extreme Environments*, London, Academic Press, pp. 297–315.
33. Horikoshi, K. and Akiba, T. (1982) *Alkalophilic Microorganisms — A New Microbial World*, Tokyo, Japan Scientific Societies Press and Berlin, Springer-Verlag.
34. Horikoshi, K., Nakao, M., Kurono, Y. and Sashihara, N. (1984) *Canadian Journal of Microbiology*, **30**, 774–9.
35. Hubalek, Z. (1976) *Folia Parasitologia*, **23**, 267–72.
36. Kakinuma, Y. (1987) *Journal of Bacteriology*, **169**, 3886–90.
37. Kakinuma, Y. (1987) *Journal of Bacteriology*, **169**, 4403–5.
38. Karlsson, S., Banhidi, Z.G. and Albertsson, A.C. (1988) *Applied Microbiology and Biotechnology*, **28**; 305–10.
39. Kashket, E.R. (1985) *Annual Review of Microbiology*, **39**, 219–42.
40. Kato, T. and Horikoshi, K. (1984) *Biotechnology and Bioengineering*, **26**, 595–8.
41. Kato, C., Honda, H., Kudo, T. and Horikoshi, K. (1984) *Journal of Fermentation Technology*, **62**, 77–80.
42. Kato, C., Kobayashi, T., Kudo, T. and Horikoshi, K. (1986) *FEMS Microbiology Letters*, **36**, 31–4.
43. Kato, C., Kudo, T., Watanabe, K. and Horikoshi, K. (1985) *Journal of General Microbiology*, **131**, 3317–24.
44. Kelly, C.T. and Fogarty, W.M. (1978) *Canadian Journal of Microbiology*, **24**, 1164–72.
45. Kelly, C.T., Brennan, P.A. and Fogarty, W.M. (1987) *Biotechnology Letters*, **9**, 125–30.
46. Khraptsova, G.I., Tsaplina, I.A., Seregina, L.M. and Loginova, L.G. (1984) *Mikrobiologiya*, **53**, 137–41.
47. Kimura, T. and Horikoshi, K. (1988) *Applied and Environmental Microbiology*, **54**, 1066–7.
48. Kimura, K., Kataoka, S., Ishii, Y., Takano, T. and Yamane, K. (1987) *Journal of Bacteriology*, **169**, 4399–402.
49. Kimura, K., Takano, T. and Yamane, K. (1987) *Applied Microbiology and Biotechnology*, **26**, 149–53.
50. Kitada, M., Lewis, R.J. and Krulwich, T.A. (1983) *Journal of Bacteriology*, **154**, 330–5.
51. Kobayashi, T., Kato, C., Kudo, T. and Horikoshi, K. (1986) *Journal of Bacteriology*, **166**, 728–32.

52. Kobayashi, H., Suzuki, T., Kinoshita, N. and Unemoto, T. (1984) *Journal of Bacteriology*, **158**, 1157–60.
53. Kobayashi, Y., Ueyama, H. and Horikoshi, K. (1982) *Agricultural Biological Chemistry*, **46**, 2139–42.
54. Koga, Y., Nishihara, M. and Morii, H. (1984) *Biochimica et Biophysica Acta*, **793**, 86–94.
55. Kogama, N., Wakabayashi, K. and Nosoh, Y. (1987) *Biochimica et Biophysica Acta*, **898**, 293–8.
56. Koyama, N., Ishikawa, Y. and Nosoh, Y. (1986) *FEMS Microbiology Letters*, **34**, 195–8.
57. Koyama, N., Niimura, Y. and Kozaki, M. (1988) *FEMS Microbiology Letters*, **49**, 123–6.
58. Krulwich, T.A. (1982) *FEMS Microbiology Letters*, **13**, 299–301.
59. Krulwich, T.A. (1966) *Journal of Membrane Biology*, **89**, 113–25.
60. Krulwich, T.A., Agus, R., Schneier, M. and Guffanti, A.A. (1985) *Journal of Bacteriology*, **162**, 768–72.
61. Krulwich, T.A., Federbush, J.G. and Guffanti, A.A. (1985) *Journal of Biological Chemistry*, **260**, 4055–8.
62. Krulwich, T.A. and Guffanti, A.A. (1986) *Methods in Enzymology*, **125**, 352–65.
63. Krulwich, T.A., Guffanti, A.A., Fong, M.Y., Falk, L. and Hicks, D.B. (1986) *Journal of Bacteriology*, **165**, 884–9.
64. Kudo, T., Ohkoshi, A. and Horikoshi, K. (1985) *Journal of General Microbiology*, **131**, 2825–30.
65. Langworthy, T.A. (1978), in D.J. Kushner (ed.) *Microbial Life in Extreme Environments*, New York and London, Academic Press, pp. 279–315.
66. Lewis, R.J., Krulwich, T.A., Reynafarje, B. and Lehninger, A.L. (1983) *Journal of Biological Chemistry*, **258**, 2109–11.
67. McLaggan, D., Selwyn, M.J. and Dawson, A.P. (1984) *FEBS Letters*, **165**, 254–8.
68. Maeda, M. and Taga, N. (1980) *Marine Ecology Progress Series*, **2**, 105–8.
69. Manachini, P.L., Fortina, M.G. and Parini, C. (1988) *Applied Microbiology and Biotechnology*, **28**, 409–13.
70. Mathrani, I.M., Boone, D.R., Mah, R.A., Fox, G.E. and Lau, P.P. (1988) *International Journal of Systematic Bacteriology*, **38**, 139–42.
71. Matsunaga, T., Nakamura, N., Tsuzaki, N. and Takeda, H. (1988) *Applied Microbiology and Biotechnology*, **28**, 373–6.
72. Mikami, Y., Miyashita, K. and Arai, T. (1982) *Journal of General Microbiology*, **128**, 1709–12.
73. Mikami, Y., Miyashita, K. and Arai, T. (1986) *Actinomycetes*, **19**, 176–91.
74. Miller, A.G., Turpin, D.H. and Canvin, D.T. (1984) *Journal of Bacteriology*, **159**, 100–6.
75. Mitchell, P. (1973) *Journal of Bioenergetics*, **4**, 63–91.
76. Miyashita, K., Mikami, Y. and Arai, T. (1984) *International Journal of Systematic Bacteriology*, **34**, 404–9.
77. Nakamura, T., Tokuda, H. and Unemoto, T. (1984) *Biochimica et Biophysica Acta*, **776**, 330–6.
78. Nishihara, M., Morii, H. and Koga, Y. (1982) *Journal of Biochemistry*, **92**, 1469–79.
79. Okazaki, W., Akiba, T., Horikoshi, K. and Akahoshi, R. (1984) *Applied Microbiology and Biotechnology*, **19**, 335–40.
80. Padan, E. and Schuldiner, S. (1986) *Methods in Enzymology*, **125**, 337–52.
81. Padan, E., Zilberstein, D. and Schuldiner, S. (1981) *Biochimica et Biophysica Acta*, **650**, 151–6.
82. Pantoliano, M.W., Ladner, R.C., Bryan, P.N., Rollence, M.L., Wood, J.F. and Poulos, T.L. (1987) *Biochemistry*, **26**, 2077–82.

83. Rosen, B.P. (1986) *Methods in Enzymology*, **125**, 328–36.
84. Sashihara, N., Kudo, T. and Horikoshi, K. (1984) *Journal of Bacteriology*, **158**, 503–6.
85. Sato, M., Arima, K. & Beppu, T. (1985) *Biotechnology Letters*, **7**, 159–64.
86. Sato, M., Beppu, T. and Aruma, K. (1983) *Agricultural Biological Chemistry*, **47**, 2019–27.
87. Schieman, D.A. (1983) *Applied and Environmental Microbiology*, **46**, 22–7.
88. Sharpe, R.J. and Munster, M.J. (1986), in R.A. Herbert and G.A. Codd (eds) *Microbes in Extreme Environments*, London, Academic Press, pp. 215–96.
89. Suhami, M., Bruyneel, B. and Verstraete, W. (1987) *Journal of Applied Bacteriology*, **63**, 117–23.
90. Takinishi, H., Sekiguchi, T., Koyama, N., Shishido, K. and Nosoh, Y. (1983) *FEMS Microbiology Letters*, **154**, 201–4.
91. Taya, M., Hinoki, H., Suzuki, Y., Yagi, T., Yap, M. and Kobayashi, T. (1985) *Journal of Fermentation Technology*, **63**, 383–7.
92. Tsai, Y., Lin, S., Li, Y., Yamasaki, M. and Tamura, G. (1986) *Biochimica et Biophysica Acta*, **883**, 439–47.
93. Tsuchida, O., Yamagata, Y., Ishizuka, T., Arai, T., Yamada, J., Takeuchi, M. and Ichisima, E. (1986) *Current Microbiology*, **14**, 7–12.
94. Ventura, S., De Phillippis, R., Materassi, R. and Balloni, W. (1988) *Archives of Microbiology*, **149**, 273–9.
95. Woese, C.R. (1987) *Microbiological Reviews*, **51**, 221–71.
96. Yamoda, H., Shimizu, S., Shimada, H., Tani, Y., Takashi, S. and Ohashi, T. (1980) *Biochimie*, **62**, 395–9.
97. Yoshihara, K. and Kobayashi, Y. (1981) *Biotechnology Letters*, **3**, 118–23.

CHAPTER 4

Oligotrophs

J.C. Fry

Introduction

Oligotrophic bacteria can be broadly defined as organisms that grow on low concentrations of organic substrates. Despite the interest shown in oligotrophs by a small band of dedicated microbial ecologists over the last 10–15 years, they are still relatively unknown to many microbiologists. In fact, even some recent texts on microbial ecology hardly mention them,[4,5,50,60] but others deal with them well.[35] It is the purpose of this chapter to offer a general introduction about these bacteria to a general microbiological audience. Most of what is known of oligotrophs has come from aquatic ecologists, so the chapter is unashamedly biased in this direction. Present knowledge indicates that true oligotrophy is a prokaryotic preserve, so eukaryotes will not be considered here.

There is a lot of variation in the terms used to describe oligotrophic bacteria. They have also been called oligocarbophilic, hypotrophic[9] and low nutrient or LN bacteria.[12] Some oligotrophs can grow only on low concentrations of carbon and these have been called obligate oligotrophs;[77] these are rare and the first pure cultures were only isolated recently.[27] Others can grow at both low and high concentrations and these have been variously called facultative oligotrophs,[25,77] euryheterotrophs[24] and eurytrophs.[9] No discussion on oligotrophic bacteria can be complete without mention of bacteria which grow at high nutrient concentrations. The different terms used for these organisms have been eutrophs,[77] heterotrophs, [1,3] saprophytes[34] and copiotrophs.[45,46]

In this chapter I will conform to the recommendations in the excellent review by Poindexter[45] and use the terms oligotroph and copiotroph; the latter is chosen because it avoids confusion with other nutritional types of bacteria. Oligotrophic bacteria will be further subdivided as facultative and obligate oligotrophs as described above. Copiotrophs that have been shown to grow only on high concentrations of organic carbon will be called obligate copiotrophs.[75] If copiotroph is used alone here, it will imply that there has been no conclusive demonstration of their absolute requirement for large amounts of carbon. Many of these organisms might in reality be facultative oligotrophs.

Habitats

Almost all fresh and marine waters contain very small amounts of dissolved organic carbon (DOC) compared with the concentrations used in conventional laboratory media. For example, most aquatic habitats contain between 1 and 6 $mgC\,l^{-1}$,[33] whereas nutrient agar contains about 4000 $mgC\,l^{-1}$. So oligotrophic environments are very widespread in nature and cannot be considered extreme habitats. Similarly, oligotrophs are probably extreme by our preconceptions rather than by their abundance.

Open ocean water contains the least organic material with DOC varying between <0.1 and 1.0 $mgC\,l^{-1}$.[40,57,76] Deep water contains less than surface water; for example, DOC values from the top 100 m and abyssal region (1000–5000 m) of the Central Pacific were 1.1 $mgC\,l^{-1}$ and 0.6 $mgC\,l^{-1}$, respectively.[13] Classically most of this carbon has been thought to be unavailable, but recent work has shown that between about 0.5 and 45% can be utilized by oligotrophic bacteria.[13] Coastal sites have higher DOC concentrations. Two examples from Japan are Osaka Bay[26] which contains 1.5–5.1 $mgC\,l^{-1}$ and Tokyo Bay[77] which contains a similar amount (2.4–3.5 $mgC\,l^{-1}$).

Freshwaters contain much higher concentrations of organic matter, the average DOC from numerous lakes in North America being 15 $mgC\,l^{-1}$.[34] The flux of DOC in lake water is, however, similar to that in seawater; values of 5 $mgC\,l^{-1}$ have been reported for eutrophic lakes, with 0.1 $mgC\,l^{-1}$ being the figure for oligotrophic lakes.[23]

Low molecular weight compounds are scarce in water; for example, seawater contains 30–50 $\mu g\,l^{-1}$.[14] Values in lake water are similar, being about 5–10 $\mu g\,l^{-1}$ for amino acids and 2–10 $\mu g\,l^{-1}$ for sugars such as glucose and sucrose.[34] Polymeric materials appear to be more abundant with peptides at about 150 $\mu g\,l^{-1}$ and total carbohydrates at 0.1–0.9 $mgC\,l^{-1}$.[3]

The amount of organic matter in water is important because it is thought to control the ratio of oligotrophs and copiotrophs. Akagi *et al.*,[3] by counts on low and high nutrient medium, showed that copiotrophs were more abundant in seawater at high carbohydrate concentrations and oligotrophs were dominant with low carbohydrate (Fig. 4.1). The dividing line occurred at 0.52 $mgC\,l^{-1}$. Surprisingly, using similar techniques, Yanagita *et al.*[77] found a great predominance of oligotrophs in the River Jinzu-gawa, which flows through a large Japanese town. They found, however, more copiotrophs in a nearby coastal water.

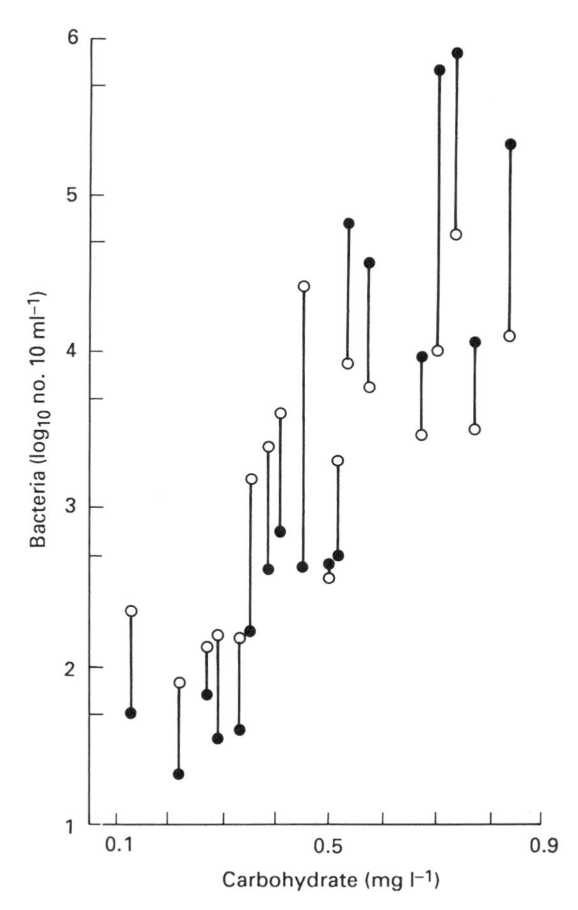

Fig. 4.1 Variation of viable counts of oligotrophic (○) and copiotrophic (●) bacteria in seawater of different total dissolved carbohydrate concentrations (from ref. 3).

The organisms

DEFINITIONS

There is no generally accepted definition of oligotrophic bacteria, however, various definitions have been suggested. Kuznetsov *et al.*[34] said oligotrophs should be defined as bacteria isolated on organic media containing between 1 and 15 mgC l^{-1}. Martin and MacLeod[38] suggested the use of 10 mgC l^{-1} and a subcommittee meeting of the Japanese Society for Microbial Ecology in 1980 recommended 1 mgC l^{-1}.[27] One group[2] recommend the production of turbidity at <10 mgC l^{-1} as a definition for oligotrophy whereas Carlucci *et al.*[12] suggest, for

Table 4.1 Classification of bacteria by their growth on agar

Organisms	*Growth*[a] *on Noble agar plates made up with*			
	Pure water[c]	*Akagi broth*[b] *containing*		
		14.8 mgC l^{-1}	*3.71 gC l*$^{-1}$	*7.42 gC l*$^{-1}$
Facultative oligotroph	+	+	+	+
Obligate oligotroph	+	+	+/−	−
Obligate copiotroph	−	−	+	+

[a]Growth (+) or no growth (−) distinguished by colony formation on agar plates viewed with a binocular microscope.
[b]Modified from ref. 3.
[c]Deionized water treated by granular activated carbon filtration and reverse osmosis.

marine isolates, their isolation on unsupplemented seawater and recommend that oligotrophs must have K_s values for growth of <10 mgC l^{-1}.

It is probable that we do not yet know enough about the physiology, abundance and distribution of either oligotrophic or copiotrophic bacteria to provide fully acceptable and precise definitions for either group.

METHODS

Agar plates

Oligotrophs have often been isolated on agar plates. Agars without any added organic nutrients have been used. In their taxonomic study Horowitz *et al.*[24] used Bushell Haas agar which has only mineral salts, and seawater agar with no mineral salts has also been used.[11,12] Others have added organic material at low concentration. Yanagita *et al.*[77] counted oligotrophs on their P media which contained 10 mg of polypeptone and either deionized water or aged seawater. Conversely, some people have used relatively rich media to isolate facultative oligotrophs. Hence, F5 agar (polypeptone, 1 g l^{-1}; yeast extract, 0.1 g l^{-1}) was reported for the isolation of two facultative bacteria.[25] Brain heart infusion estuarine salts agar (brain heart infusion, 3.7 g l^{-1}; mineral salts, 15 g l^{-1}) was used by MacDonell and Hood[36] to isolate very small aquatic bacteria, 89% of which were oligotrophic. Most of these methods involved incubating the plates for 5–28 days at 5–23 °C.

Plating on agar is often used to distinguish oligotrophs from copiotrophs. For example, Yanagita *et al.*[77] defined oligotrophs as those that grew on P medium but not on R medium (polypeptone, 10 g l^{-1}; yeast extract, 5 g l^{-1}). This approach is not always reliable because many facultative oligotrophs will also grow on rich media. So it would be better to use a suite of media on which organisms can be replica plated and classified accordingly. At Cardiff[75] we have developed such an approach (Table 4.1). Although we originally used Akagi broth[3] for this test, it also works well with R2A medium[49] on which oligotrophs grow better.

Glass fibre filters

Akagi *et al.*[3] developed a method for the isolation and enumeration of oligotrophs which avoided agar, with its uncertain complement of utilizable organic materials, but retained the advantages of a solid surface. This method incubated the aquatic bacteria, concentrated on polycarbonate membrane filters (47 mm in diameter, 0.2 µm pore size), on heated (450–500 °C) glass fibre filters (70 mm) soaked in growth medium. The dilute nutrient medium used contained 16.8 mgC l^{-1} and has been used by many others since. The medium contained (as mgC l^{-1} in charcoal treated seawater): polypeptone, 10; proteose peptone, 5; Bacto-soytone, 5; yeast extract, 5; sodium glycollate, 5; sodium malate, 5; D-mannitol, 5; sucrose, 5; and ferric citrate, 0.5. Some other studies have successfully used this procedure as well.[2,17]

Dilution to extinction in liquid media

Classical most probable number (MPN) enumeration methodology has also been used to enumerate oligotrophs, both in pure culture[27] and in nature.[26] Both methods diluted the organisms in low nutrient medium and then incubated them at 20 °C for 2–4 weeks. Ishida and Kadota[27] used LT10^{-4} medium (trypticase, 0.5 mgC l^{-1}; yeast extract, 0.05 mgC l^{-1}; in aged seawater), whereas in their 1979 study[26] they used a variety of media based on the TFl recipe (trypticase, 5 gC l^{-1}; yeast extract, 0.5 gC l^{-1}; in artificial lake water) at concentrations down to 10^{-4} of TF1. They also used filtered lake water alone. The problem with MPN methods in dilute media is that growth is hard to detect. So in both studies they added ^{14}C-labelled substrates (glutamate, glucose or protein hydrolysates) and detected growth by $^{14}CO_2$ evolution using a scintillation counter. In addition, the 1979 study[26] detected growth by plating on TF1 and TF10^{-1} agars.

Baxter and Sieburth[9] used dilution series of inorganically supplemented seawater with 0.01 and 0.1 mgC l^{-1} of glucose, incubated at 17 °C for 2 days. They detected growth by epifluorescence microscopy and used dilutions with both negative and positive tubes to inoculate another series. After three or more series they subcultured onto agar to obtain isolates; disappointingly only facultative oligotrophs were isolated by this elaborate procedure.

Enrichment isolation

Chemostat enrichments have been used by some to isolate bacteria capable of growth at low nutrient concentrations. Jannasch[28] used a sterile seawater medium supplemented with very low concentrations of sodium lactate or glucose (0.036–10.8 mgC l^{-1}) at dilution rates down to 0.01 h^{-1}. Higher concentrations of carbon were used by Matin and Veldkamp,[39] who isolated a *Spirillum* sp. on 0.5 g l^{-1} of lactate at 28 °C with a dilution rate of 0.05 h^{-1}. The organisms isolated by these procedures have relatively high K_s values (see later) compared with bacteria isolated in other ways. This observation supports Poindexter's[45] assertion that chemostat enrichment is not as good as plating methods for oligotroph isolation. It has also been suggested that bacteria with the adaptations for oligotrophy are only isolated in enrichments with less than 10 mg l^{-1} of organic material.[10]

Table 4.2 Some oligotrophs that have been identified, their habitat and nutritional type

Bacterium	*Source or habitat*	*Type*[a]	*Reference*
Spirillum sp.	Freshwater pond	—	39
Pseudomonas sp. 486	Coastal water, Japan	F	1
Pseudomonas sp. RP-303	Coastal water, Japan	F	2
Arthrobacter spp.	Soil	—	
Caulobacter cresentus	Freshwater and marine	—	45
Asticcacaulis biprosthecum	—	—	
Aeromonas sp. No. 6	Lake Biwa, Japan	F	25
Flavobacterium sp. M1	Lake Mergozzo, Italy	F	
Pseudomonas fluorescens P17	Drinking water	F	73
Flavobacterium sp. S12	Tap water	F	67
Spirillum sp. NOX	Slow sand filter	F	68
Hyphomicrobium sp. T37	Freshwater	—	47
Acinetobacter sp. GO1	Seawater	—	9
Agromonas sp.	Soil, rice roots	—	41
Corynebacterium sp. MC2	Canal water, UK	F	
Curtobacterium sp. CF2	Taff feeder canal, UK	F	32
Pseudomonas fluorescens MD5	Spring water, UK	F	
Bacillus pumilis WFO1	Llanishen reservoir, UK	F	75
Pseudomonas sp. WOO1	Llanishen reservoir, UK	O	

[a]F = facultative oligotroph; O = obligate oligotroph; — = unknown

Van der Kooij and Hijnen[66,68,69] have used batch culture enrichments to isolate facultative oligotrophs from drinking water. They incubated tap water with added substrates at 15 °C. *Flavobacterium* spp. were isolated with added starch, strain 166 on 10–20 $\mu g C l^{-1}$ and strain S12 on 100 $\mu g C l^{-1}$. Similarly, a *Spirillum* sp. NOX was isolated on a mixture of formate, glycollate and oxalate, all at 25 $\mu g C l^{-1}$. As these isolates grew well on low concentrations of one or more of the substrates used in the enrichment, this method might have general application for isolating oligotrophs with specific nutrient requirements.

SPECIES FOUND

Unfortunately many of the oligotrophic bacteria that have been isolated and studied in some detail have not been identified. Table 4.2 gives a list of some of the oligotrophs that have been identified. Most are facultative oligotrophs, probably because more of this type have been isolated and they grow on the richer nutrient media commonly used for identification tests. Obligate oligotrophs present severe identification problems. Ishida and Kadota[27] described all of the eighteen obligate oligotrophs that they isolated as non-motile, Gram-negative, catalase and oxidase positive rods. Although three were curved rods, none could be identified. West and Fry[75] identified one of the obligate oligotrophs they isolated as *Pseudomonas* sp. However, this identification was far from certain because, although the majority of tests indicated this genus, the strain would not grow on nutrient agar, on which all *Pseudomonas* spp. are supposed to grow by definition (M. Rhodes-Roberts, personal communication).

It is clear from Table 4.2 that oligotrophs are from varied genera and sources. Several of the genera are prosthecate bacteria and most of the others are *Pseudomonas*, *Flavobacterium* or *Spirillum*. Few are Gram-positive bacteria. There have been other studies and reviews of oligotrophic bacterial species. Kuznetsov *et al.*[34] lists many of the genera in Table 4.2 but also include *Agrobacterium*, *Photobacterium*, *Vibrio*, *Micrococcus*, *Staphylococcus*, *Microcyclus*, *Leptothrix*, *Ochrobium*, *Metallogenium* and *Pasteuria*. Akagi *et al.*[2] performed a taxonomic study on oligotrophs and copiotrophs isolated from three marine sites on the Japanese coast. They found little taxonomic difference between the two nutritional groups, but most of the oligotrophs were in the groups *Pseudomonas*, *Vibrio* and *Acinetobacter–Moraxella*.

An oligotroph classification system was developed[75] (Table 4.1) to investigate 421 isolates from a reservoir. In this study most were found to be facultative (94.8%) but eight (1.9%) were obligate oligotrophs and fourteen (3.3%) were obligate copiotrophs. These percentage occurrences of oligotrophs were similar to Ishida and Kadota's[26] results, from a nutrient rich site in Lake Biwa. These scientists showed that 0.9–2.4% of the viable bacteria were oligotrophs; however, they also found many more (5.3–66.7%) from a nutrient poor site in the same lake.

West and Fry[75] also surveyed forty pure cultures of laboratory bacteria and showed most to be facultative oligotrophs (95%); none were obligate oligotrophs. The facultative bacteria were from the genera *Acinetobacter*, *Bacillus*, *Beneckia*, *Corynebacterium*, *Escherichia*, *Klebsiella*, *Micrococcus*, *Proteus*, *Providencia*, *Pseudomonas*, *Serratia*, *Staphylococcus* and *Streptococcus*. Many of these would normally be considered as copiotrophs and have not been identified as oligotrophs before. However, Yanagita *et al.*[77] did find that one of their laboratory strains of *Escherichia coli* was facultatively oligotrophic; they also report an obligately oligotrophic mutant derived from *E. coli* W3110.

STABILITY OF OLIGOTROPHIC CHARACTERISTICS

Most publications make no mention of the stability of the oligotrophs used and so presumably found them stable. However, sometimes oligotroph instability has been specifically mentioned. Akagi *et al.*[2] report that some lost their capacity to grow on low nutrient medium. Other bacteria isolated under low nutrient conditions have shown increases in K_s values (between 1.7- and 12-fold) after repeated subculture on nutrient rich medium ($c\,0.7\,\mathrm{gC\,l^{-1}}$).[28] MacDonell and Hood[36] studied the growth profiles of two oligotrophs, in trypticase broth. One strain, UM106, hardly grew in $2\,\mathrm{g\,l^{-1}}$ on initial isolation, but, after forty days subculture on brain heart infusion estuarine salts agar, it produced maximal optical densities at this concentration. Thus one organism was stable and one unstable.

In my laboratory in Cardiff we have had experience in working with oligotrophs for about three years. Only one facultative oligotroph, *Bacillus pumilis*

WFO1, has proved unstable, by completely losing its capacity to grow on low nutrient media. All the other obligate and facultative oligotrophs handled have shown very stable growth profiles on media of different carbon concentrations. This has been true whether they have been maintained on high or low nutrient media. The copiotrophs have equal stability.

Adaptations for oligotrophy

SUBSTRATE UTILIZATION

It has been suggested that oligotrophic bacteria should be able to take up and grow on a wide variety of different carbon compounds to take advantage of whatever substrates become available in low nutrient environments.[45] This suggestion appears to be true for many of the organisms that have been studied. In fact, Horowitz *et al.*[24] found bacteria isolated on low nutrient media to be more tolerant to variations in salinity and pH, and to be nutritionally less fastidious than bacterial populations isolated on rich media.

Oligotrophs identified as *Pseudomonas* spp. seem to have particularly broad substrate utilization profiles. *P. fluorescens* P17 grew well on 73% of forty-five substrates at $2.5\,g\,l^{-1}$ in agar[71] and on 72% of a similar range of twenty-nine substrates at $1\,gC\,l^{-1}$ in agar.[75] In similar tests using thirty-eight substrates, from a slightly broader range, the same bacterium only grew on 66% of the substrates; this was similar to the 63% recorded for another *P. fluorescens* MD5 and both showed much broader profiles than the other oligotrophs tested (Kemmy and Fry, unpublished). A *Pseudomonas* sp. WOO1 that was obligately oligotrophic had a similar wide substrate range on agar, growing well on 69% of twenty-nine substrates provided at $1\,gC\,l^{-1}$.[75]

Some other oligotrophs are similar; *Aeromonas* sp. No. 6 utilized twenty-one sugars and amino acids from the twenty-nine tested (72%). An unidentified oligotroph, 486, grew on all the eight substrates tested by Martin and MacLeod[38] at $10\,mgC\,l^{-1}$, but on only five at $1\,gC\,l^{-1}$. The same authors also tested two copiotrophs, RP-303 and RP-250; the former grew on all eight substrates at the high concentration but on only three low concentration substrates, whereas the latter grew on all substrates at both concentrations. These results indicate that RP-303 and RP-250 are facultative oligotrophs, which supports my previous suggestion that many copiotrophs are not solely copiotrophic.

Other oligotrophs have narrower substrate profiles. For example, *Flavobacterium* sp. S12 grew significantly on only twelve carbohydrates and glycerol from sixty four natural organic compounds provided at $1\,g\,l^{-1}$ on agar.[67] When these sixty-four substrates were provided at low concentration ($10\,\mu gC\,l^{-1}$) in liquid medium S12 grew only on sucrose, maltose, raffinose, starch and glycerol. However, the bacterium also grew on maltotriose, -tetraose, -pentaose, -hexaose and stachyose. Clearly many of these substrates are structurally related to the starch on which it was originally isolated. My group at Cardiff have found similar narrow substrate profiles for other facultative oligotrophs. For example, a *Curtobacterium* sp. grew on only seven of thirty-eight substrates. It grew on only the complex materials, like peptone and yeast extract, and three carboxylic acids but not on any amino acids or carbohydrates.

Ishida and Kadota[27] tested the eighteen obligate oligotrophs they isolated from Lake Biwa by incubation with ^{14}C-organic substrates at low concentration in $LT10^{-4}$ liquid medium. They found that all organisms utilized glutamic acid, glycine, serine and glycollate, for which they had a requirement. However, only one used glucose and none utilized acetate, proline and leucine.

Many oligotrophs will grow in either pure water or mineral salts media carefully prepared to be as free from organic matter as possible[27] (Fig. 4.4a). This growth might be at the expense of organic materials contaminating these media, as contaminants of the mineral salts or from the atmosphere, perhaps as vapours of volatile organics that are plentiful in laboratories. Another possibility is that successful oligotrophs can fix carbon dioxide either heterotrophically or autotrophically. Experiments with ^{14}C-labelled bicarbonate have shown that some oligotrophs can fix small amounts of carbon dioxide but that this accounts for only a small proportion of the carbon utilized (West and Fry, unpublished). Yield coefficients well above 1.0 have been obtained with two oligotrophic *Pseudomonas* spp. below 1 $mgC\,l^{-1}$ and these could not be explained by carbon dioxide fixation (West and Fry, unpublished). This evidence suggests that contaminating organic materials may account for the growth observed in some laboratory experiments at extremely low concentrations of added carbon.

GROWTH YIELDS

The prime adaptation of oligotrophs must be their ability to grow at very low concentrations of organic carbon. To study this many people have measured the growth yield in liquid media at different concentrations of carbon. Akagi *et al.*[2] used optical density as the index of growth on polypeptone in artificial seawater. The oligotrophs they isolated all produced perceptible turbidity in 10 $mgC\,l^{-1}$, whereas the copiotrophs did not. Some isolates (e.g. O-81 and O-59; Fig. 4.2a) gave almost linear responses between 10 $mgC\,l^{-1}$ and 5 $gC\,l^{-1}$ and so were clearly facultative oligotrophs. One isolate, 490 (Fig. 4.2b) did not grow above 1 $gC\,l^{-1}$ and only gave yields similar to O-81 and O-59 up to 100 $mgC\,l^{-1}$. Thus 490 appeared to be an obligate oligotroph. Martin and MacLeod[38] used some of the strains from Akagi *et al.*[2] to examine the effect of the type of substrate on their growth profiles. The substrate used greatly affected the growth profile (see also previous section). For example, RP-250 grew to almost ten times the optical density on glutamate and succinate than on peptone at carbon concentrations below about 2 $gC\,l^{-1}$ (Fig. 4.2c). Thus oligotrophs can make use of some substrates far better than others.

Optical density measurements are not ideal for estimating the growth yields of oligotrophs because the very low carbon concentrations that support growth give small absolute yields. So some researchers have used viable counting methods. Obligate oligotrophs were studied by counting with a MPN method, using $LT10^{-4}$ medium with ^{14}C-glutamate uptake used as the growth indicator in the tubes.[27] The results (Fig. 4.3) showed clearly that none of the bacteria could grow

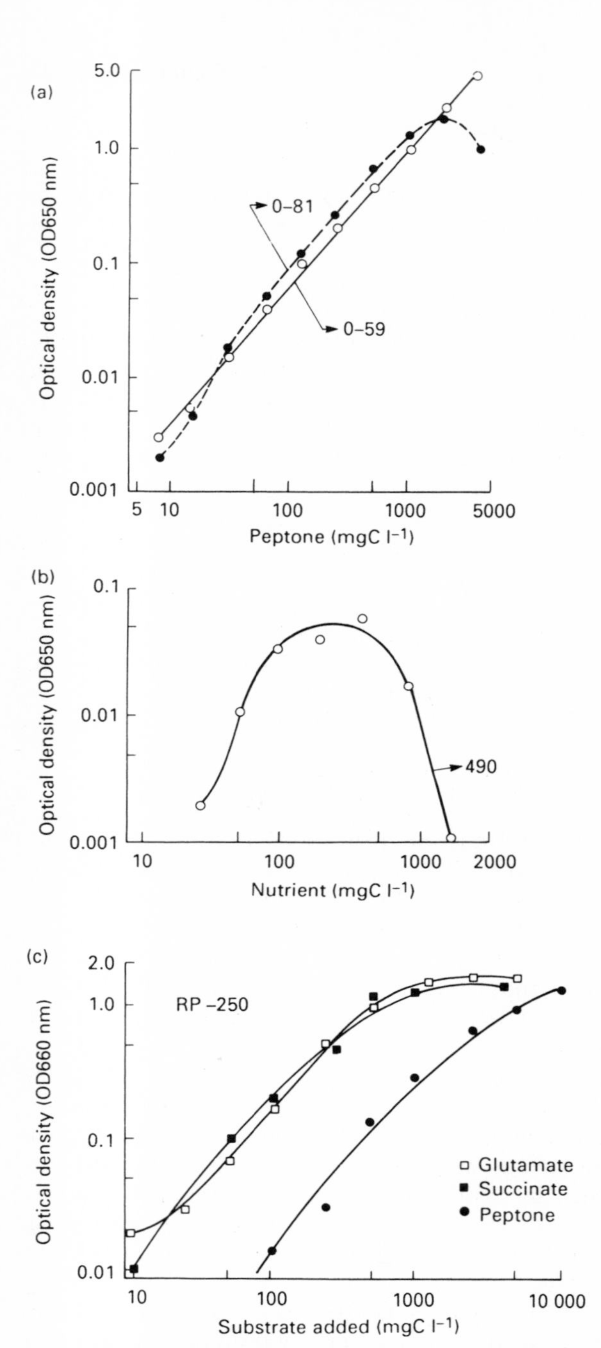

Fig. 4.2 Growth profiles of bacteria obtained by optical density on various concentrations of organic nutrients. (a) Two facultative oligotrophs (strains O-81 and O-59). (b) One obligate oligotroph (strain 490). (c) A facultative oligotroph (RP-250) with carbon added as peptone (●), succinate (■) or glutamate (□). Results in (a) and (b) from ref. 2 and in (c) from ref. 38.

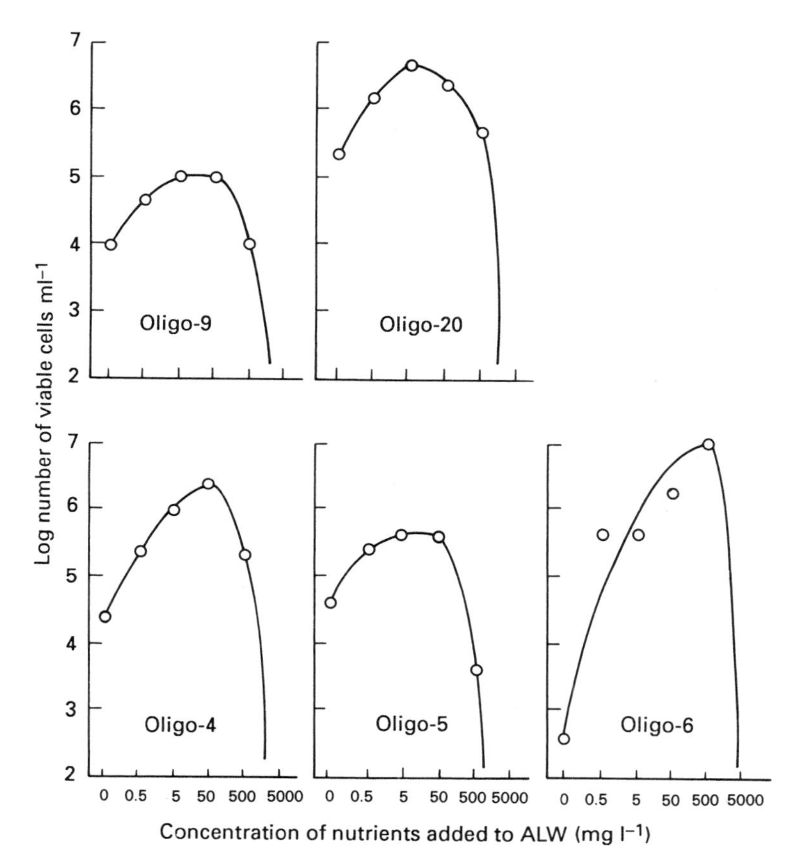

Fig. 4.3 Growth profiles of five obligately oligotrophic bacteria on various concentrations of organic nutrients, using most probable number counts with ^{14}C-glutamate uptake as the growth indicator (from ref. 27). ALW = aged lake water.

above about $1\,gC\,l^{-1}$ and that maximum yields were about 10^5–$10^7\,ml^{-1}$ at 50–1000 $mgC\,l^{-1}$. It was also clear that even with no added carbon bacterial yields of about 3×10^2 to $3 \times 10^5\,ml^{-1}$ were obtained. Work in my laboratory has confirmed and extended these observations (Fig. 4.4a).[75] An obligate oligotroph, *Pseudomonas* sp. WOO1, did not grow above $6\,gC\,l^{-1}$, whereas the facultative *P. fluorescens* P17 gave high yields right up to $74\,gC\,l^{-1}$ but failed to grow at $234\,gC\,l^{-1}$. Both organisms attained similar populations at the very low substrate concentrations. A copiotroph was also studied in these experiments and had a similar growth profile to P17 at high carbon but failed to grow at all below $10\,mgC\,l^{-1}$. These studies showed that the medium used affected the success of obligate as well as facultative oligotrophs, because WOO1 grew far better in R2A broth than in Akagi broth, especially above about $100\,mgC\,l^{-1}$.

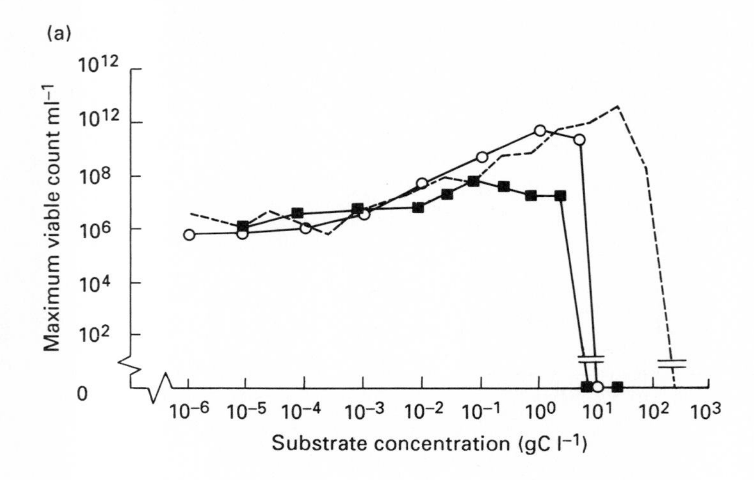

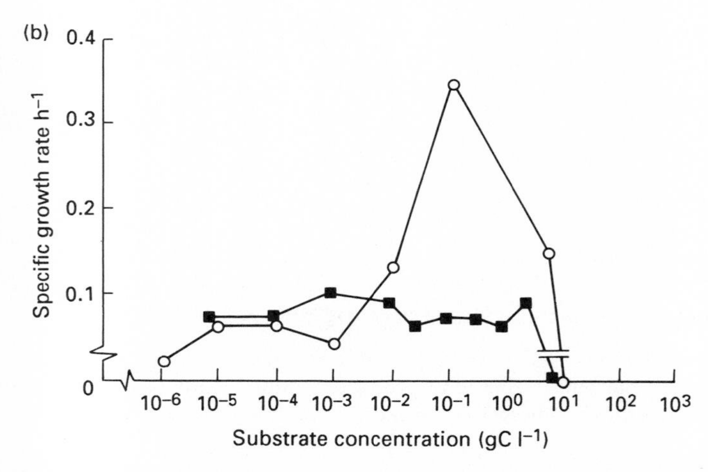

Fig. 4.4 Growth responses of an obligate oligotroph *Pseudomonas* sp. WOO1 (○, ■) and a facultative oligotroph *P. fluorescens* P17 (– – –) using (a) maximum viable counts (from ref. 75) and (b) specific growth rate (West and Fry, unpublished). The experiments were done on Akagi (■) and R2A (○) broths. Traditional plate counts were used as the growth indicator.

GROWTH KINETICS

To be able to grow effectively on low concentrations of substrates oligotrophs would be expected to have high affinity uptake systems and so low saturation constants for growth (K_s values). Many of those studied do indeed have such low K_s values. Carlucci *et al.*[12] have found some of the lowest saturation constants reported; the values were lowest on single substrates but also low on peptone and yeast extract (Table 4.3).

Table 4.3 K_s values for oligotrophic bacteria isolated from open ocean water (from ref. 12)

	K_s (μgC l^{-1}) *for isolate*		
Substrates	*150*	*151*	*169*
Peptone and yeast extract	4.9	3.3	3.8
Glucose	0.8	—	1.4
Acetate	0.4	0.2	0.2
Proline	3.0	3.0	—

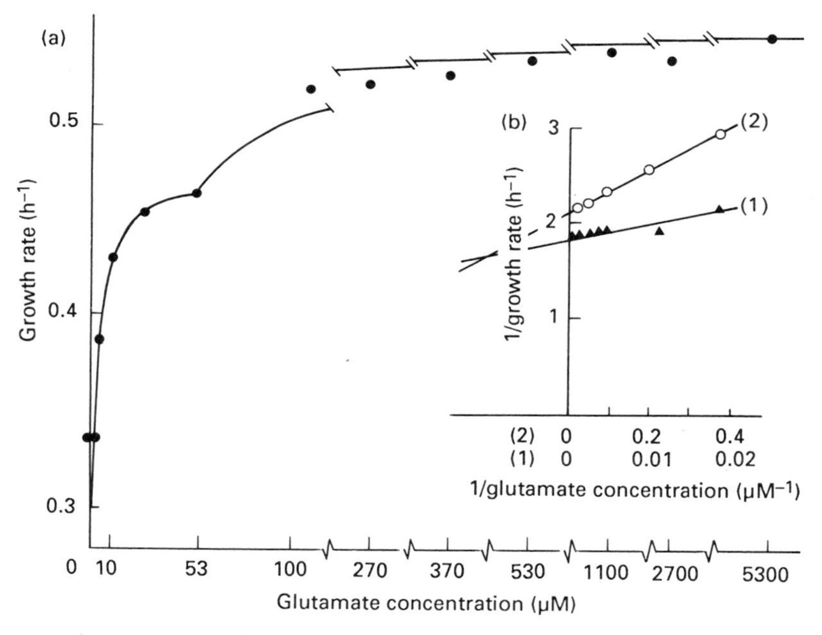

Fig. 4.5 (a) Growth rate of a facultative oligotroph *Aeromonas* sp. No. 6 as a function of glutamate concentration. The curve was determined by linear regression analyses of the Monod equation. (b) Reciprocal plots of the growth rate against glutamate concentration (from ref. 25).

Many bacteria have both low affinity and high affinity enzyme systems.[63] This means that a plot of growth rate against substrate concentration will be biphasic (Fig. 4.5) or even multiphasic,[75] having two or three saturation curves following each other. Table 4.4 gives a summary of the K_s values of five oligotrophic and two copiotrophic bacteria with multiple affinity enzyme systems. The obligate oligotroph *Pseudomonas* sp. WOO1 had the lowest constant closely followed by the facultative *P. fluorescens* P17, which is clearly almost as efficient at growth on low nutrient concentrations. These K_s values were very similar to those of the marine oligotrophs discussed above (Table 4.3).[12]

Table 4.4 K_s for some bacteria[a] with both high and low affinity enzyme systems at different substrate concentrations

Bacterium	*Substrate*	K_s (μgC l^{-1})	*Substrate concentration* (μgC l^{-1})	Reference
Pseudomonas sp. WOO1	R2A broth	1.5	1–1000	75
		12 200	>1000	
P. fluorescens P17	Akagi broth	2.5	7–1320	
		1587	1320–131 000	
		60 800	>131 000	
Flavobacterium sp. M1	Glucose	12.6	252–1000	25
		648	1000–19 800	
		60 800	>19 800	
Aeromonas sp. No. 6	Glutamate	70	160–3140	
		530	>3140	
Asticcacaulis biprosthecum	Glucose	130	—	45
		2450	—	
Escherichia coli 7020^{+}	Glucose	8460	12 420–198 000	25
		46 800	>198 000	
Flavobacterium sp. M2[b]	Glucose	28 000	12 420–198 000	
		1 314 000	>198 000	

[a]Placed in order of minimum substrate affinity
[b]Copiotrophic bacteria

Although the lowest K_s for *Flavobacterium* sp. M1 was 12.6 μgC l^{-1} it is very close to the suggested minimum value for an oligotroph (10 μgC l^{-1}).[12] However, the *Aeromonas* and the prosthecate, *Asticcacaulis*, have much higher minimum values of around 100 μgC l^{-1}. The equivalent values for the two remaining bacteria in Table 4.4, which are copiotrophs,[25] are about two orders of magnitude higher, about 10 000 μgC l^{-1}.

Thus there is a clear relationship between the minimum K_s and the nutritional type of organism. This conclusion is further supported by other results.[75] A rather poor facultative oligotroph that reverted to copiotrophy during routine handling, *Bacillus pumilis* WFOl, had a minimum K_s value of 235 μgCl^{-1}. This value is clearly above those for all the oligotrophs in Table 4.4. The one obligate copiotroph we have values for had a minimum K_s of 110 000 μgC l^{-1}, which was well above the equivalent values for the two copiotrophs in Table 4.4. Perhaps there is a possibility for a sliding scale of oligotrophy and copiotrophy, based on minimum K_s, to be used in the future, rather than the more rigid definitions suggested so far.

Saturation constants for growth can be estimated only when there are significant differences between growth rates, so clearly K_s values could not be estimated for *Pseudomonas* sp. WOO1 on Akagi broth (Fig. 4.4b). This has clearly also been a problem for some of the bacteria in Table 4.4, because the lowest substrate concentration which was used to calculate K_s is sometimes rather high.

Table 4.5 K_s values for the facultative oligotroph *Flavobacterium* sp. S12 on various substrates (from ref. 69)

Substrate	K_s(μgC l^{-1})
Maltotriose	5.7
Maltose	23.7
Glucose	109.2
Starch	8.4
Amylose	25.6
Amylopectin	11.0

As has already been discussed above, from substrate utilization and growth profiles, some oligotrophs behave differently on different substrates. Saturation constants for growth also vary, as might be expected. An example is given in Table 4.5, where the K_s values for *Flavobacterium* sp. S12 were sufficiently different for it to have been classified as an oligotroph on some substrates but not on others, with respect to the criterion of Carlucci *et al.*[12]

Several bacteria that have been called oligotrophs by some workers are clearly borderline candidates according to the sliding scale proposed above. For example, the chemostat enriched *Spirillum* sp. isolated by Matin and Veldkamp[39] had a K_s value of 1310 μgCl^{-1}, which puts it right between the oligotrophs and copiotrophs. Their *Pseudomonas* sp., which was out-competed by the *Spirillum* sp. at low nutrient concentration, falls in the same category.

The results described above (Table 4.4; Fig. 4.5) indicate very clearly that oligotrophs can adapt their substrate affinity according to the nutrient concentration that they are growing in. It also appears, from the shapes of the kinetic curves (Fig. 4.5), that they do this in a step-wise manner, turning off one enzyme system as they utilize another. This might be achieved by substrate inhibition, as Bell[10] shows that concentrations higher than 10 mg l^{-1} favour the development of populations without such adaptations.

UPTAKE KINETICS

Although there has been less research on the uptake kinetics of oligotrophs, the results that have been obtained are very similar in principle to those discussed above for growth kinetics. So a detailed discussion of this topic will not be given. Thus oligotrophs have low and high affinity uptake systems[25] and the minimum K_m values for uptake for oligotrophs are smaller than for copiotrophs.[1,25,63] There is also similar variability between substrates; for example, proline uptake is more efficient than glucose uptake in oligotrophs 486 and RP-303.[1]

OTHER ADAPTATIONS

Several other adaptations for oligotrophy probably exist but have been far less studied than the factors discussed above. Although some of these adaptations have been discussed previously,[45] a few will be briefly mentioned here.

One would expect a high surface area to volume ratio for efficient nutrient uptake. This is assured by some obligate oligotrophs that do not increase in size when they grow on high substrate concentrations.[27] This is unlike most other bacteria that become smaller as substrate concentration, and so growth rate, decreases.[63] Prosthecate bacteria have another mechanism to increase the cell surface area for more efficient uptake: this is to lengthen the prosthecum. This is exactly what these organisms do in low nutrient medium, when the length of the prosthecum can be five to forty times the length of the cell.[45,46] There is strong evidence that the prosthecum helps accumulate organic nutrients, as in *A. biprosthecum* the intra-organelle concentration can be 200-fold the ambient value. It has also been suggested that the primary role of the prosthecum is the uptake of low phosphate concentrations and that organic oligotrophy is a consequence of this property.[47]

The maintenance energy requirement is rarely studied in oligotrophic bacteria. However, Matin and Veldkamp[39] have shown that their *Spirillum* sp. had a maintenance requirement only 25% that of the *Pseudomonas* sp. that it outcompeted in the chemostat. A low energy of maintenance implies a high survival potential. Unfortunately, few studies have been done on the survival of oligotrophs. However, it is well known that aquatic bacteria, many of which are small and probably oligotrophic,[36] survive very well at low nutrient concentrations.[42,58] It has also been suggested that bacteria might survive in water by entering a viable but non-culturable state;[15] oligotrophs might adopt this strategy.

Commercial aspects

The specific properties of oligotrophs have not yet been harnessed in any commercial production process, although some bacteria used in industry might, incidentally, be facultative oligotrophs. The only commercial use to which oligotrophs have been put is for the bioassay of assimilable organic carbon (AOC) in drinking waters. Bacteria cause regrowth problems in water mains,[18] which can lead to excessive bacterial slime production and invertebrate infestations. Thus water engineers need to keep bacterial regrowth to a minimum. As bacterial growth is limited by carbon concentrations, it has been suggested that mains waters with low concentrations of bacterially utilizable DOC will have less regrowth potential. Classical methods of measuring the amount of organic matter in water, as DOC or total organic carbon, do not distinguish between biologically unavailable carbon, such as humic acids, and AOC. Thus, if predictions of bacterial regrowth are to be made from a knowledge of the amount of carbon in the water, it is the AOC which should be measured. If the AOC is measured, then

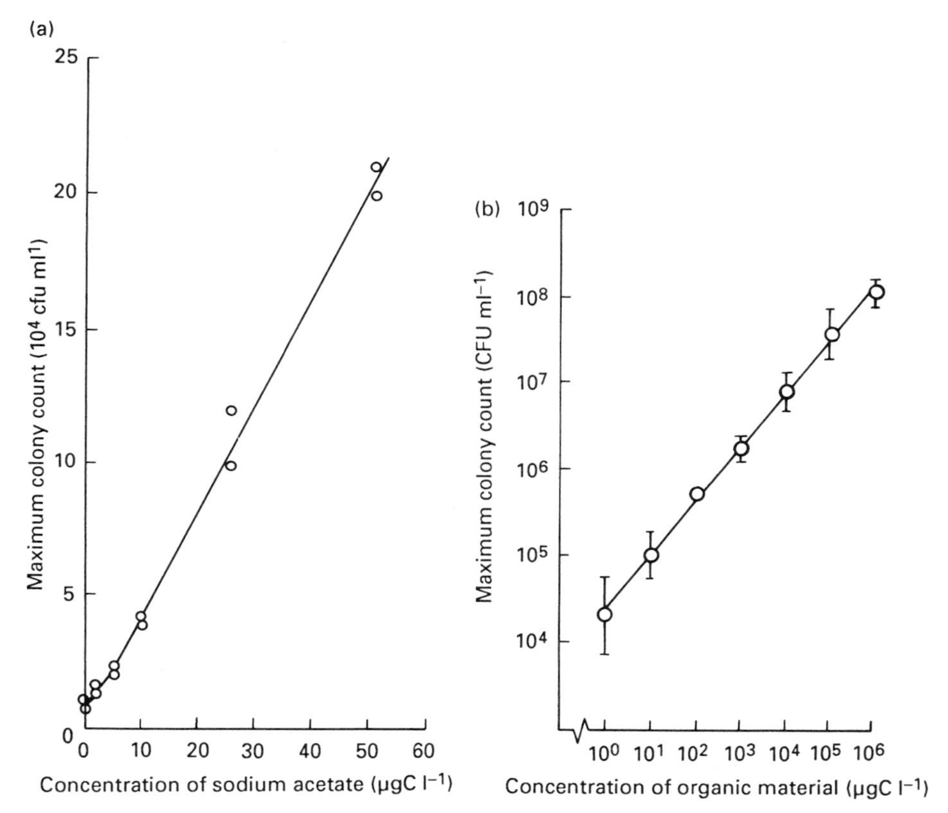

Fig. 4.6 Calibration curves for two methods of bioassay for estimating the amount of assimilable organic carbon in drinking water. (a) Relationship between maximum colony counts of *Pseudomonas fluorescens* P17 and different concentrations of sodium acetate added to tap water (from ref. 73). (b) Relationship between colony counts after 6 days incubation of a mixture of four pure cultures of facultatively oligotrophic bacteria and different concentrations of organic material added to pure water. Results are means of nine calibration curves $\pm 95\%$ confidence intervals (from ref. 32).

treatment methods could be redesigned to minimize AOC and hence bacterial regrowth.

So far six different approaches for estimating the concentration of AOC in potable waters have been suggested in the literature. All the methods are by necessity bioassays because bacteria grow on a heterogeneous and as yet undefined fraction of the organic carbon in water which cannot be assayed chemically. Four of these approaches use mixed natural populations as inocula in the bioassay,[31,44,56,74] so will not be discussed here. The other two approaches both use pure cultures of facultative oligotrophs as inocula.

Van der Kooij and his group working in The Netherlands Waterworks Testing and Research Institute have developed one approach. Their method involved growing single pure cultures of bacteria, originally isolated from drinking water, at 15 °C without shaking on pasteurized water samples (60 °C for 30 min). The growth of the bacteria was estimated daily by plate counting until the maximum count was reached, which was usually within 8 days. The maximum count was then converted to an estimate of AOC by comparison with a calibration curve obtained in the same way from tap water amended with single carbon compounds such as acetate, starch, and oxalate. Several different bacteria have been used for this bioassay. *P. fluorescens* P17 was used[73] because it grew on a wide range of carbon compounds. Also used was *Flavobacterium* sp. S12 because it grew well on polysaccharides[68] and *Spirillum* sp. NOX which grew well on oxalate.[66,69] Assays using these bacteria were very sensitive with the published calibration curves giving straight lines between 0 and 11 to 100 μgCl^{-1} (e.g. Fig. 4.6a). In another case, using *Aeromonas hydrophila* 315, a calibration curve between 3 μgCl^{-1} and 1×10^4 μgCl^{-1} showed linearity above 10 μgCl^{-1} and only a slight deviation below this concentration.[72] Thus measurement of AOC over a very wide range of concentrations is clearly possible with this technique.

The other technique[32] used a mixture of four bacteria as inoculum in the bioassay. The organisms used were *Pseudomonas fluorescens* MD5, *Corynebacterium* sp. MC2, *Curtobacterium* sp. CF2 and an unidentified coryneform MR2. They were all isolated by enrichments on drinking water without added substrates with inocula from surface waters. Water samples were dechlorinated with 0.02% (w/v) sodium thiosulphate before inoculation and the indigenous bacteria were removed by membrane filtration. The four bacteria were then inoculated, at 100 colony forming units ml^{-1}, in 50 ml samples of the water to be tested at 20 °C for 6 days in static culture. After incubation the growth yields of the bacteria were estimated by a drop counting procedure on nutrient agar. A growth curve was not needed in this bioassay; this made the method better suited to use in a routine water industry laboratory than the previous method. The yield of bacteria was converted to AOC by reference to a calibration curve obtained by similarly incubating the bacteria in pure water containing a mixture of organic carbon compounds at concentrations between 0 and 1×10^6 μgCl^{-1} (Fig. 4.6b).

Using a mixture of facultative oligotrophs has many advantages over the single organism approach. It provides a more robust assay where the mixture of organisms provided are more likely to be able to utilize all the available carbon compounds in the water than a single organism (see section on substrate utilization). Recently Van der Kooij and his team[65,70] have modified their method by using both *P. fluorescens* P17 and *Spirillum* sp. NOX in the inoculum.

These bioassays for AOC have already proved very useful to water engineers in The Netherlands who have used the results to improve treatment processes.[65,70] Also some water treatments have been found to increase AOC and so should be used with caution; one example of this is ozonation.[29,68]

Genetic aspects

There have been no specific genetic studies on oligotrophs. There is, however, a wealth of genetic information on bacteria such as *E. coli*, *Acinetobacter* spp. and *Pseudomonas* spp., some strains of which could be oligotrophic. In this section I shall not attempt to cover these species. Instead, I shall provide a brief summary about what is known of the genetics of bacteria in natural waters, which are the major oligotrophic habitats.

Many reviews have been written which suggest the potential for genetic exchange in water[48,61,64] However, few experiments have been done specifically *in situ* or on specially isolated aquatic bacteria. Most studies have been with conjugal transfer systems, but a few reports on transduction and transformation are also present in the literature.

In 1984 Gowland and Slater[22] reported the conjugal transfer of drug resistance plasmids between *E. coli* strains in dialysis sacs suspended in a pond. However, only recently have experiments been done without enclosure, so the bacteria are exposed to the natural, competing microflora. Bale *et al.*[6,8] have shown that one mercury resistance, conjugal plasmid (pQM1) isolated from a river transferred between strains of *Pseudomonas aeruginosa* in unenclosed experiments *in situ* at a high rate ($<2\times10^{-2}$ per recipient). This transfer was highly dependent on both temperature and nutrient concentration. There was a linear relationship between river water temperature and the $\log_{10}$ of the transfer frequency over the range 6–21 °C. The optimum temperature for the *in situ* transfer of plasmid-encoded mercury resistance between the natural, epilithic bacterial population and a *P. putida* strain was about 20 °C.[7]

More studies have been carried out on bacteria isolated from aquatic habitats specifically for genetic work. Gauthier *et al.*[19] report that eight out of thirty-one strains of marine bacteria could transfer mercury resistance to *E. coli* in complex media at frequencies between 10^{-3} and 10^{-8}. A more recent study[21] examined the ability of sixty-eight freshwater bacteria to receive and express two broad host range plasmids from *P. aeruginosa*. The conjugative plasmid R68 was transferred to 38% of the isolates at frequencies between about 10^{-2} and 10^{-6}. Mobilization of a small, non-self-transmissible plasmid (R1162) by R68 was also observed in 15% of these freshwater strains. Some studies have studied conjugal plasmid transfer with aquatic isolates in more detail. Temperature profiles for transfer of three plasmids isolated specifically from aquatic bacteria showed optima between 10 and 25 °C.[53] This study also showed an optimum of 25 °C for transfer from a mixed natural suspension of epilithic bacteria to *P. aeruginosa*. Thus these results are broadly supportive of the *in situ* observations described previously. Cell density, nutrient concentration and donor: recipient ratio also affected plasmid transfer with both natural isolates and mixed natural suspensions from freshwater, but pH had much less of an effect.[53] However, with marine isolates transfer of mercury resistance occurred best at 30 °C and salt was essential in the mating medium.[19] Although many of these natural plasmids appear stable in their bacterial hosts, one mercury resistance plasmid (pQM3) underwent substantial rearrangement during transfer with some hosts.[51] Despite this, analysis of natural plasmids from aquatic bacteria with restriction endonucleases suggests that structural rearrangements might occur in the environment.[7,30,53]

In 1978 Morrison *et al.*[43] demonstrated that chromosomal genes of *P. aeruginosa* could be transferred by the generalized transducing phage F116 in flow-through test chambers in a freshwater reservoir. Similar work using phage ΦDS1, has shown transduction of plasmid Rms149 between non-lysogenic and lysogenic strains of *P. aeruginosa*.[55] This transfer occurred both in the presence and the absence of the natural microflora, used natural concentrations of bacteria and was markedly affected by the donor: recipient ratio. In these experiments transfer frequencies varied between about 10^{-5} and 10^{-2} for the chromosomal genes and between 10^{-8} and 10^{-5} for the plasmid.

Transformation experiments have not yet been carried out in natural aquatic habitats. However, a small, mercury resistance plasmid (pQM17) isolated from epilithic strains of *Acinetobacter calcoaceticus* has been studied in the laboratory.[52] This plasmid transferred best when whole cells of both donor and recipient were present. This process is called natural transformation[59] and was DNAase sensitive. The temperature optimum for the process was very broad (14–37 °C) and it was also affected by the concentration of organic nutrients and the donor: recipient ratio, but not by pH.

There has been a lot of interest recently in the transfer, spread and survival of genetically engineered microorganisms (GEMs) in the environment.[62] However, most studies have been done in terrestrial habitats and little is known of this subject in water. One substituted benzoate degrading GEM, *Pseudomonas* sp. B13 strain FR1, containing the recombinant plasmid pFRC20P, has been investigated in a laboratory chemostat fed with synthetic sewage and containing activated sludge.[16] The bacterium survived for 14 days at a population density of about 10^5 per ml but no plasmid transfer was detected. Such transfer would appear to be possible for several reasons. The nonconjugative, genetically engineered plasmids pBR322 and pBR325 can transfer by mobilization from *E. coli* to natural wastewater strains.[20] These plasmids sometimes formed cointegrates which could perhaps be transferred to other natural bacteria by the mechanisms discussed earlier. Other studies in laboratory microcosms of both wastewater treatment plants and river epilithon have confirmed that transfer by mobilization or direct conjugation can occur under almost natural conditions.[37,54]

These results suggest that plasmid transfer between oligotrophic bacteria in natural habitats is a real possibility. We await specific genetic studies on oligotrophic isolates for confirmation.

Future prospects

There is clearly a need for a lot more basic research on the physiology, occurrence and distribution of oligotrophs. The research which has been done on the physiology of these organisms can be regarded only as a start. Most is known about how oligotrophs grow on organic substrates. Much more biochemical work is needed on the mechanisms used for uptake and growth at very low concentrations of substrate. There is also a need for basic research on the maintenance requirement, turnover of macromolecules and survival of oligotrophs. There are enough stable strains of both obligate and facultative oligotrophs for this biochemical work to be done. I hope suitably experienced biochemists and microbiologists will be sufficiently motivated by reading this chapter and the associated references to take up this challenge. It should not be left to microbial ecologists alone to study these fascinating bacteria.

Genetic work is more difficult because suitable strains and genetic markers have not yet been developed for oligotrophs. There might be hope for the development of genetic systems in the known facultative oligotrophs, because their growth on the richer laboratory media might make phenotypic selection comparatively easy. Obligate oligotrophs that only grow on dilute media will almost certainly be much more difficult to examine genetically.

Not until a great deal more is known about the basic physiology of oligotrophs will their commercial potential be realized. It might well be that a whole new group of bacteria with a wide range of biosynthetic capabilities is waiting to be discovered.

Summary

Two types of oligotrophs can be identified. Obligate oligotrophs are comparatively rare in nature and can only grow on low concentrations of carbon (<1–$6\,gC\,l^{-1}$). Facultative oligotrophs are common in nature and can grow at both very low and high concentrations of carbon. Many commonly used laboratory bacteria are facultative oligotrophs. Most oligotrophs so far identified are prosthecate bacteria or belong to the genera *Pseudomonas* or *Flavobacterium*. Oligotrophs tend to utilize a broad spectrum of carbon compounds. The occurrence of two or three low and high affinity uptake systems for organic substrates is common in these bacteria. The minimum K_s value for growth indicates the oligotrophic potential of the bacterium, with low values (*c*1–150 $\mu gC\,l^{-1}$) clearly indicating oligotrophy. However, these minimum K_s values were fairly evenly spread between 1.5 and 110 000 $\mu gC\,l^{-1}$ for an obligate oligotroph and an obligate copiotroph, respectively. This made it very difficult to make clear demarcations between these two physiological groups. Other adaptations for oligotrophy might include small size, prosthecum formation and a low maintenance requirement. Facultative oligotrophs are used in bioassays for assimilable organic carbon in water treatment and distribution systems, but they have no other commercial use at present. Almost nothing is known about the genetics of oligotrophs. There is a clear need for more research on the physiology and biochemistry of these fascinating organisms.

Acknowledgements

I would like to thank Miss Alison West, Miss Frances Kemmy and Mr Julian Diaper for the research they have done under my supervision into various aspects of oligotrophy. I also thank Miss West for the many useful discussions we have had about oligotrophs and oligotrophy; without these this review would have been much poorer. I am also grateful to my collaborator Dr M.J. Day, and past research workers Dr M.J. Bale and Dr P.A. Rochelle for help with the genetic research reported here. I thank Severn–Trent Water and the Natural Environment Research Council for providing support for the research on oligotrophs and ecological genetics carried out in my laboratory.

References

1. Akagi, Y. and Taga, N. (1980) *Canadian Journal of Microbiology*, **26**, 454–9.
2. Akagi, Y., Simidu, U. and Taga, N. (1980) *Canadian Journal of Microbiology*, **26**, 800–6.
3. Akagi, Y., Taga, N. and Simidu, U. (1977) *Journal of Microbiology*, **23**, 981–7.
4. Atlas, R.M. and Bartha, R. (1987) *Microbial Ecology*, Menlo Park, California, USA, The Benjamin/Cummin Publishing Company.
5. Austin, B. (1988) *Marine Microbiology*, Cambridge University Press.
6. Bale, M.J., Fry, J.C. and Day, M.J. (1987) *Journal of General Microbiology*, **133**, 3099–107.
7. Bale, M.J., Fry, J.C. and Day, M.J. (1988) *Applied and Environmental Microbiology*, **54**, 972–8.
8. Bale, M.J., Day, M.J. and Fry, J.C. (1988) *Applied and Environmental Microbiology*, **54**, 2756–8.
9. Baxter, M. and Sieburth, J.McN. (1984) *Applied and Environmental Microbiology*, **47**, 31–8.
10. Bell, W.H. (1984) *Microbial Ecology*, **10**, 217–30.
11. Carlucci, A.F. and Shimp, S.L. (1974), in R.R. Colwell and R.Y. Morita (eds) *Effect of the Ocean Environment on Microbial Activities*, Baltimore, University Park Press, pp. 363–7.
12. Carlucci, A.F., Shimp, S.L. and Craven, D.B. (1986) *FEMS Microbiology Ecology*, **38**, 1–10.
13. Carlucci, A.F., Shimp, S.L. and Craven, D.B. (1987) *FEMS Microbiology Ecology*, **45**, 211–20.
14. Chau, Y.K. and Riley, J.P. (1966) *Deep Sea Research*, **13**, 115–24.
15. Colwell, R.R., Brayton, P.R., Grimes, D.J., Roszak, D.B., Huq, S.A. and Palmer, L.M. (1985) *Biotechnology*, **3**, 817–20.
16. Dwyer, D.F., Rojo, F. and Timmis, K.N. (1988), in M. Sussman, C.H. Collins, F.A. Skinner and D.E. Stewart-Tull (eds) *The Release of Genetically Engineered Micro-organisms*, London, Academic Press, pp. 77–88.
17. Emiliani, F. (1984) *Hydrobiologia*, **111**, 31–6.
18. Fransolet, G., Villiers, G. and Masschelein, W.J. (1985) *Ozone Science and Engineering*, **17**, 205–25.
19. Gauthier, M.J., Cauvin, F. and Briettmayer, J. (1985) *Applied and Environmental Microbiology*, **50**, 38–40.
20. Gealt, M.A., Chai, M.D., Alpert, K.B. and Boyer, J.C. (1985) *Applied and Environmental Microbiology*, **49**, 836–41.
21. Genthner, F.J., Chatterjee, P., Barkay, T. and Bourquin, A.W. (1988) *Applied and Environmental Microbiology*, **54**, 115–17.
22. Gowland, P.C. and Slater, J.H. (1984) *Microbial Ecology*, **10**, 1–13.
23. Hood, D.W. (1970) *Organic Matter in Natural Waters*, Alaska, Institute of Marine Science, University of Alaska.

24. Horowitz, A., Krichevsky, M.I. and Atlas, R.M. (1983) *Canadian Journal of Microbiology*, **29**, 527–35.
25. Ishida, Y., Imai, I., Miyagaki, T. and Hajime, K. (1982) *Microbial Ecology*, **8**, 23–32.
26. Ishida, Y. and Kadota, H. (1979), in A.W. Bouquin and P.H. Pritchard (eds), *Microbial Degradation of Pollutants in Marine Environments*, Florida, US Environmental Protection Agency, pp. 135–47.
27. Ishida, Y. and Kadota, H. (1981) *Microbial Ecology*, **7**, 123–30.
28. Jannasch, H.W. (1968) *Journal of Bacteriology*, **95**, 722–3.
29. Janssens, J.G., Meheus, J. and Dirickx, J. (1984) *Water Science and Technology*, **17**, 1055–68.
30. Jobling, M.G., Peters, S.E. and Ritchie, D.A. (1988) *FEMS Microbiology Letters*, **49**, 31–7.
31. Joret, J.C. and Levi, Y. (1986) *Tribunale Cedbedeau*, **39**, 3–9.
32. Kemmy, F.A., Fry, J.C. and Breach, R.A. 1989. *Water Science and Technology*, **21**, 155–9.
33. Kushner, D.J. (1978), in D.J. Kushner (ed.) *Microbial Life in Extreme Environments*, London, Academic Press, pp. 1–7.
34. Kuznetsov, S.I., Dubinina, G.A. and Lapteva, N.A. (1979) *Annual Review of Microbiology*, **33**, 377–87.
35. Lynch, J.M. and Hobbie, J.E. (1988) *Micro-organisms in Action: Concepts and Applications in Microbial Ecology*, Oxford, Blackwell Scientific.
36. MacDonell, M.T. and Hood, M.A. (1982) *Applied and Environmental Microbiology*, **43**, 566–71.
37. Mancini, P., Fertels, S., Nave, D. and Gealt, M.A. (1987) *Applied and Environmental Microbiology*, **53**, 665–71.
38. Martin, P. and MacLeod, R.A. (1984) *Applied and Environmental Microbiology*, **47**, 1017–22.
39. Matin, A. and Veldkamp, H. (1978) *Journal of General Microbiology*, **105**, 187–97.
40. Menzel, D.W. and Ryther, J.H. (1968) *Deep Sea Research*, **15**, 327–37.
41. Mochizuki, M. and Hattori, T. (1986) *FEMS Microbiology Ecology*, **38**, 51–55.
42. Morita, R.Y. (1982) *Advance in Microbial Ecology*, **6**, 171–98.
43. Morrison, W.D., Miller, R.V. and Sayler, G.S. (1978) *Applied and Environmental Microbiology*, **36**, 724–30.
44. Pike, E.B., Stanfield, G., Jago, P.H. and Irving, T.E. (1987) *Proceedings of the American Water Works Association Conference on Water Quality Technology*, Denver, Colorado, American Water Works Association, 607–18.
45. Poindexter, J.S. (1981) *Advances in Microbial Ecology*, **5**, 63–89.
46. Poindexter, J.S. (1981) *Microbiological Reviews*, **45**, 123–79.
47. Poindexter, J.S. (1984), in M.J. Klug and C.A. Reddy (eds), *Current Perspectives in Microbial Ecology*, Washington DC, American Society for Microbiology, pp. 33–40.
48. Reanney, D.C., Gowland, P.C. and Slater, J.H. (1983), in J.H. Slater, R. Wittenburg and J.W.T. Wimpenny (eds), *Microbes in their Natural Environments*, Cambridge University Press, pp. 379–421.
49. Reasoner, D.J. and Geldreich, E.E. (1985) *Applied and Environmental Microbiology*, **49**, 1–7.
50. Rheinheimer, G. (1985) *Aquatic Microbiology*, Chichester, John Wiley & Sons.
51. Rochelle, P.A., Day, M.J. and Fry, J.C. (1988) *FEMS Microbiology Letters*, **52**, 245–50.
52. Rochelle, P.A., Day, M.J. and Fry, J.C. (1988) *Journal of General Microbiology*, **134**, 2930–41.
53. Rochelle, P.A., Fry, J.C. and Day, M.J. (1989) *Journal of General Microbiology*, **135**, 409–24.

54. Rochelle, P.A., Fry, J.C. and Day, M.J. (1989) *FEMS Microbiology Ecology*, **62**, 127–36.
55. Saye, D.J., Ogunseitan, O., Sayler, G.S. and Miller, R.V. (1987) *Applied and Environmental Microbiology*, **53**, 987–95.
56. Servais, P., Billen, G. and Hascoet, M.C. (1987) *Water Research*, **21**, 445–50.
57. Sharp, J.H. (1975), in T.M. Church (ed.) *Marine Chemistry in the Coastal Environment*, Washington DC, American Chemical Society, pp. 682–96.
58. Stevenson, L.H. (1978) *Microbial Ecology*, **4**, 127–33.
59. Stewart, G.J., Carlson, C.A. and Ingraham, J.L. (1983) *Journal of Bacteriology*, **156**, 30–5.
60. Stolp, H. (1988) *Microbial Ecology: Organisms, Habitats, Activities*, Cambridge University Press.
61. Stotzky, G. and Babich, H. (1986) *Advances in Applied Microbiology*, **31**, 93–138.
62. Sussman, M., Collins, C.H., Skinner, F.A. and Stewart-Tull, D.E. (1988) *The Release of Genetically Engineered Micro-organisms*, London, Academic Press.
63. Tempest, D.W. and Neijssel, O.M. (1978) *Advances in Microbial Ecology*, **2**, 105–53.
64. Trevors, J.T., Barkay, T. and Bourquin, A.W. (1987) *Canadian Journal of Microbiology*, **33**, 191–8.
65. Van der Kooij, D. (1986) The effect of treatment on assimilable organic carbon. In *Treatment of Drinking Water for Organic Contaminants*, P.M. Huck and P. Toft (eds), New York, Pergamon Press, pp. 317–28.
66. Van der Kooij, D. and Hijnen, W.A.M. (1981) *Applied and Environmental Microbiology*, **41**, 216–21.
67. Van der Kooij, D. and Hijnen, W.A.M. (1983) *Applied and Environmental Microbiology*, **45**, 804–10.
68. Van der Kooij, D. and Hijnen, W.A.M. (1984) *Applied and Environmental Microbiology*, **47**, 551–9.
69. Van der Kooij, D. and Hijnen, W.A.M. (1985) *Applied and Environmental Microbiology*, **49**, 765–71.
70. Van der Kooij, D., Hijnen, W.A.M. and Kruithof, J.C. (1987) *Proceedings of the 8th Ozone World Congress of the International Ozone Association*, **1**, D96–D113.
71. Van der Kooij, D., Oranje, J.P. and Hijnen, W.A.M. (1982) *Applied and Environmental Microbiology*, **44**, 1086–95.
72. Van der Kooij, D., Visser, A. and Hijnen, W.A.M. (1980) *Applied and Environmental Microbiology*, **39**, 1198–204.
73. Van der Kooij, D., Viser, A. and Hijnen, W.A.M. (1982) *Journal of the American Water Works Association*, **74**, 540–5.
74. Werner, P. (1985) *Vom wasser*, **65**, 258–70.
75. West, A.J. and Fry, J.C. (1989) *Applied and Environmental Microbiology* (submitted).
76. Williams, P.M., Carlucci, A.F. and Olson, R. (1980) *Oceanologica Acta*, **3**, 471–6.
77. Yanagita, T., Ichikawa, T., Tsuji, T., Kamata, Y., Ito, K. and Sasaki, M. (1978) *Journal of General Applied Microbiology*, **24**, 59–88.

CHAPTER 5

Osmophiles

D.H. Jennings

Introduction

Microorganisms which are able to grow under conditions in which water is not readily available have been described as osmophilic.[83] Since then synonyms have been coined such as osmotophilic,[82] osmotolerant,[2] osmoduric,[82] osmotrophic,[70] xerophilic[64] and xerotolerant.[11] As we shall see, it could be argued that osmotolerant might be the best term for these microorganisms but for the moment we shall use the original name, osmophiles. Before describing the organisms themselves, we need to say something about how availability of water to an organism is defined.

CONCEPTUAL FRAMEWORK

When a cell grows it increases in volume, part of that increase being brought about by an increasing volume of water within the outer membrane. The rate of flow of water into the cell must be such that the pressure generated does not cause the membrane to burst. When the cell has a wall, the pressure which the outer covering can stand is much greater. Indeed the generation of a significant hydrostatic pressure within the cell is part of the driving force for wall extension and therefore cell expansion.

The driving force for movement of water into a cell is the difference in its chemical potential across the outer membrane. Water moves down the gradient of *chemical potential*. Essentially, there is a difference in free energy of the water on either side of the membrane. In plant physiology, the term *water potential* is used to give a measure of the chemical potential of water within a system.[23] However it must be remembered that, although water potential can be considered equivalent to chemical potential, it is not formally so. In thermodynamic terms, the water potential of a system is the chemical potential divided by the partial molar volume.

The magnitude of the water potential (and indeed chemical potential of water) will depend on the forces acting on the water in the system which do not exist in pure water under standard conditions. If a significant pressure is present, the water molecules will be pushed slightly closer together and therefore their free energy will be increased. Therefore the chemical potential and the water potential will be raised above unity and have a positive value. Likewise, if solutes are dissolved in the water, the solvent molecules will be less reactive so that the water potential will decrease and will be negative. In plant physiology, it is customary to separate out, in a consideration of water potential, the forces within the bulk phase of the solution which affect the chemical potential of water from those at surfaces, which at least in physico-chemical terms may be different, from those in the bulk phase. In strict thermodynamic terms, this is not correct but in conceptual terms it can be useful as I shall show below. This particular matter is nicely discussed elsewhere.[9] Be that as it may, water molecules may be either repelled or attracted to surfaces. Thus their free energy may be increased or decreased. In this way the surface may lead to the water potential being increased (the forces may make the value more positive) or decreased (the value may be made more negative).

From the above, the water potential of a system — and here we can think of a walled cell — is given by

$$\Psi = \Psi_p - \Psi_s \pm \Psi_m$$

where Ψ is the water potential of the cell; Ψ_p, the pressure or *turgor potential*, that component of Ψ brought about by the pressure generated by the relatively inelastic wall; Ψ_s, the *solute potential*, that contribution due to dissolved solutes; Ψ_m, that contribution due to surfaces or interfaces, namely the *matric potential*.

In the cell, the matric potential for nearly all purposes can be ignored. Thus for the walled cells we have

$$\Psi = \Psi_p - \Psi_s$$

But there are situations, particularly for microbes, because of their size, where matric forces are significant. Soil is one very good example. This will be so not only for liquid–solid interfaces but also for liquid–gas interfaces.

In microbiology, it has been customary to use *water activity* (a_w) as a measure of the ability of a system to gain or lose water; $-a_w$ being the vapour pressure above the solution, medium or organism (p) relative to the standard state (p_o). At constant temperature, a_w is logarithmically related to water potential. This is because the chemical potential of water ($\bar{\mu}$) is related to vapour pressure as follows:

$$\bar{\mu} = RT \log_e p/p_o = RT \log_e a_w$$

Therefore

$$\Psi = RT \log_e a_w / V$$

where R is the gas constant ($J\,mol^{-1}\,K^{-1}$); T the temperature (K); V the partial molar volume. Such a description of water availability has been unattractive to

Table 5.1 Values for water potential (Ψ) and equivalent values for water activity (a_w) and the molality of sodium chloride solutions which would produce these values at 25 °C. Data from refs 17, 32, 86

Ψ (−MPa)	a_w	NaCl molality (mol [kg water]$^{-1}$)
1.38	0.99	0.3
2.78	0.98	0.607
5.62	0.96	1.2
8.52	0.94	1.77
11.5	0.92	2.31
14.5	0.90	2.83
17.6	0.88	3.32
20.8	0.86	3.80
30.7	0.80	5.15
39.1	0.753	6.16

For reference: 1.032 M NaCl has a molality of 1.054 mol [kg water]$^{-1}$

those dealing with eukaryotic microbes because of a concern, when considering the dynamics of water movement into or out of a cell, to know how the individual components of the water potential, namely Ψ_p and Ψ_s, are changing. Table 5.1 gives the relationship between water activity and water potential in numerical terms.

The habitats

It is not difficult to see that there could be many environments where osmophilic microorganisms might be found. Perhaps the best approach is to follow the lead of Griffin[32] who classified xeric environments according to the contribution of the matric and solute potential to the water potential. Table 5.2 indicates some of the major environments. For Griffin,[32] by far the most important division is between group A and the others. In group A unicellular organisms predominate, whereas in groups B and C it is the filamentous fungi. Griffin[30,33,89,90] believes that the reduced significance of unicellular organisms in structures (soils) of low matric potential is due to the restriction of movement, regardless of whether it is brought about by the activity of the cell itself or by Brownian motion. In the simplest situation, with the cell on a surface, for movement to occur, a film of water of at least the diameter of the cell must be present, or it or its progeny will be trapped at one site by surface tension. If one is considering pores within the matrix, for movement to occur the pores must be water-filled and must have a diameter of the appropriate size. It is not difficult to calculate that, for a bacterial cell of 1 μm, no pore with a radius large enough to allow movement of that cell will remain water-filled at a matric potential of − 147 kPa. This value compares with − 100 kPa, the value for the soil water potential at which activities of soil bacteria are very severely reduced. Griffin argues that filamentous fungi are able to spread because

Table 5.2 A classification of environments inhabited by osmophilic microorganisms and the characteristic genera and species there found (from ref. 32)

	A		*B*	*C*	
Group	*a*	*b*		*a*	*b*
Major constituent	Liquid		Porous solid	Porous solid	
Solute potential	Low		High	Low	
Matric potential	High		Low	Low	Variable
Nutrient concentration	Low	High	Low	Low	High
Examples	Salt lake	Syrups: preservatives brines	Soil of low water content	Saline soil of low water content	Stored food
Characteristic microorganisms	Bacteria and algae (*Halobacterium*, *Halococcus*, *Dunaliella*)	Yeasts (*Debaryomyces hansenii*, *Zygosaccharomyces rouxii*)	Filamentous fungi (*Aspergillus*, *Penicillium*)	Filamentous fungi (*Aspergillus*, *Penicillium*)	Filamentous fungi (*Aspergillus*, *Penicillium*, *Chrysosporium*, *Eremascus*, *Monascus*, *Wallemia*)

hyphal extension is still possible even through water films that are thinner than the diameter of the hyphae.

This is a somewhat simplistic view of the situation within a soil and assumes that a water potential preventing bacterial activity is due primarily to the matric potential restricting movement and not any other cause. In any case, even when the matric potential prevents movement of bacteria, it will no more prevent the spread of a bacterial colony than it does a fungal colony, even though for the same doubling time the former colony will spread inherently more slowly than the latter, because the internal forces producing cell growth are essentially the same as for hyphal extension. However, while the colony growth will account for some fungal spread through soil, it is likely that the more effective means of spread will be by spore dispersal, almost certainly at times when water is readily available. Even in soils which are dry for long periods, there will also be a flux of water at some time due to rain. In this respect it needs to be remembered that for fungal spores or cells with a radius of 5 μm, appropriately sized pores in the soil will no longer be water-filled with Ψ_m at a value of only -14.7 kPa. Thus for spread by gravity, Brownian movement or cell motility, bacterial cells will be increasingly at a greater advantage than fungal cells as a soil dries out from field capacity. Finally, if matric forces were the major factor limiting growth of bacteria in soils, it is hard to see why they are much less likely to be found in many xeric environments where matric forces are small. This matter of the much greater sensitivity of bacteria than fungi to low water potentials will be considered further at a later stage.

When we consider group A (Table 5.2), the division into two saline groups is a consequence of the input of combined carbon into the system. In natural saline systems, such carbon enters via photosynthesis and therefore in such environments photosynthetic cells will predominate. In the most saline natural environments, such as salt lakes, one finds highly specialized organisms such as *Halobacterium* and *Dunaliella*. Although osmophiles, they are more suitably described as halophiles because the cells need to cope not only with the very low external water potential but also with the very high external concentration of sodium chloride. The physiology of these organisms is well discussed in Chapter 6. Here it is sufficient to say that the bacteria can absorb sodium chloride and possess enzymes which are adapted to the high concentrations of the salts which are generated within the cells. In the case of *Dunaliella*, though there is some adaptation of the enzymes, sodium chloride is to a considerable degree excluded from the metabolic machinery of the cell, much of the necessary internal solute potential being generated by the production of glycerol by photosynthesis.

In the artificial saline media, in which light is excluded, it is the yeasts which grow at the lowest medium water potential. Though some information about the physiology of these yeasts is presented below, the information is nowhere near as full as one would want. Thus, whereas glycerol is known to make an important contribution to the internal solute potential of the cells when they grow in saline media, the concentration of ions within the cells can be relatively high. Yet we do not know whether the enzymes are adapted to these high concentrations or whether the ions are sequestered mainly in the large vacuole which is characteristic of these yeasts.

In all other habitats listed in Table 5.2 it is the fungi which become increasingly significant with declining water potential. This is in keeping with what has been found in the laboratory studies referred to above. At the lowest water potentials, it is the filamentous fungi which predominate. The reason why these fungi are not found in preserving syrups and brines is a consequence of the low oxygen tensions in these media and of the moulds being strict aerobes. On the other hand, many of the osmophilic yeasts can ferment vigorously under anaerobic conditions. However, this is not true of all such yeasts; *Debaryomyces hansenii* grows particularly poorly in the absence of oxygen.[76]

The organisms

CLASSIFICATION

In the case of higher organisms, their very size allows their ready classification on an ecological basis, since bulk environmental features can be used to define the ecological amplitude of the species under consideration. Thus a higher plant can be classified readily, e.g. as xerophyte, halophyte, etc. Studies in the laboratory are carried out not to check the veracity of any classification but to understand the mechanisms which allow a species to grow in particular ecological situations. On the other hand, with microorganisms such classification is not always possible, since one cannot be certain that the isolated organism is actively growing within the environment or living in a niche in which the environmental factors are different from those in the environment at large. Because of this uncertainty and because of the ease with which microorganisms can be grown in the laboratory, there have been many studies on the effect of putative environmental factors on growth of selected microorganisms in order to determine their ecological characteristics. In particular, there have been many studies concerned with the effect of different water potentials on growth. From such studies, microbes can be put into groups according to the range of water potentials that can be withstood.

Before indicating the classification which can be made on the above lines, some comments are necessary on the procedures used. In a medium where there are no matric forces, such that water potential equates with the solute potential, i.e. the microorganism is grown on agar or in liquid, account must be taken of the solute used to change the medium solute potential. The ability of the cells to absorb the solute will have an effect on the growth rate which is observed. If the solute can be transported into the cell rapidly by an active process, such that the concentration inside is greater than outside, cell growth will be greatly assisted because turgor can be readily generated. There might be problems if the solute is toxic to the cell when accumulated to higher than normal concentrations. This can be a possibility if sodium chloride is used to change the medium solute potential.

The relationship of solute absorption to medium water potential is that described by Oertli with respect to elongation of higher plant cells in media of different salinities.[61] He postulated that solute requirement for cell elongation at constant turgor increases linearly with decreasing external Ψ brought about by increasing the concentration of solute. Solute absorption must be such that it obeys

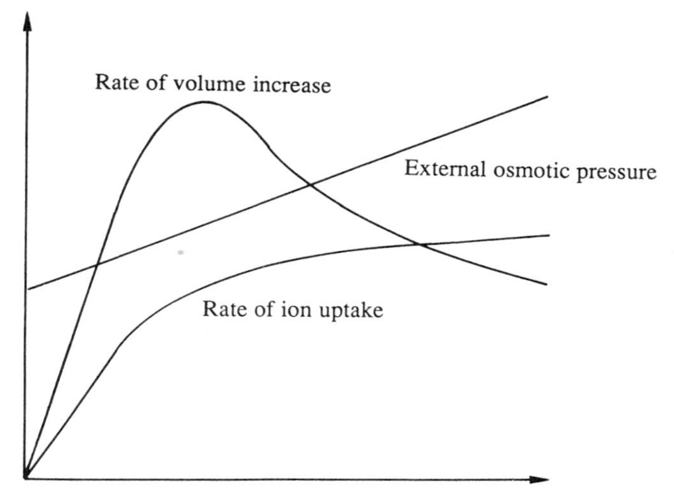

Fig. 5.1 The relationship between the solute potential within a eukaryotic walled organism, solute uptake and growth as measured by increase in volume of the organism when it is growing in a range of external water (=solute) potentials. (Based on ref. 61).

saturation kinetics with respect to Ψ, taking it to be a measure of solute concentration. Therefore, at near zero solute supply, a finite requirement has to be fulfilled with virtually no solutes and thus the elongation rate approaches zero. At the other extreme, an infinite rate must be satisfied with a supply which is finite because of the rate of functioning of the transport system. In the latter case the elongation rate must again decrease towards zero. The situation is exemplified in Fig. 5.1.

It follows from the above, that in order to assess how water availability *per se* might be affecting growth, there is a need to use several solutes to change the external Ψ_s. Ideally it would help if one could find a solute which was not absorbed by the organism. With the majority of microbes this is well nigh impossible, given the adaptability of their uptake systems and their ability to secrete degradative enzymes.

In all the above, it has been assumed that the matric potential contributes little to the medium water potential. As will be evident in the next section, it will be argued that for a substrate such as soil, the matric potential can make a major contribution to the water potential to which a microorganism might be exposed. A number of methods have been proposed whereby the matric potential of material such as soil might be altered.[15,20,48] However, it should be noted that it is by no means clear that the changes in water potential produced by such methods are due solely to changes in matric potential, as indeed might be surmised by considerations set out in the previous section.

Table 5.3 Approximate values for the lowest water potentials (Ψ) allowing growth of algae, fungi, bacteria and actinomycetes in culture. (Information obtained from refs 11, 17, 18, 31, 58, 65, 78, 80)

Group	Ψ *(MPa)*	*Algae*	*Fungi*	*Bacteria*	*Actinomycetes*
1	−2	Freshwater species *Chlorella emersonii* *Scenedesmus obliquus*	Many wood decay species *Serpula lacrimans*	*Mycoplasma gallisepticum*	
2	−5	Brackish and marine species	Many phycomycetous and coprophilous species *Phytophthora cinnamomi* *Pythium ultimum*	Many Gram-negative species *Pseudomonas* spp. Enterobacteriaceae	Many soil species
3	−5 to −20		Soil fungi such as *Alternaria alternata* *Cladosporium cladosporoides* *Fusarium culmorium* *Rhizoctonia solani* *Saccharomyces cerevisiae*	Many Micrococcaceae *Staphylococcus aureus* Most *Bacillus* spp. Many Lactobacillaceae *Clostridium botulinum* Moderate halophiles *Paracoccus (Micrococcus) halodenitrificans* *Vibrio parahaemolyticus*	Many soil species
4	−20 to −40		Many *Aspergillus*, *Eurotium* and *Penicillium* spp. Osmophilic yeasts on salt *Debaryomyces hansenii* *Saccharomyces rouxii*		
5	below −40	*Dunaliella parva* and *D. viridis*	*Chrysosporium fastidium* *Monascus bisporus* Some isolates of *Aspergillus restrictus* *Zygosaccharomyces rouxii* on sugars	Halophilic bacteria *Halobacterium halobium*	*Actinospora halophila*

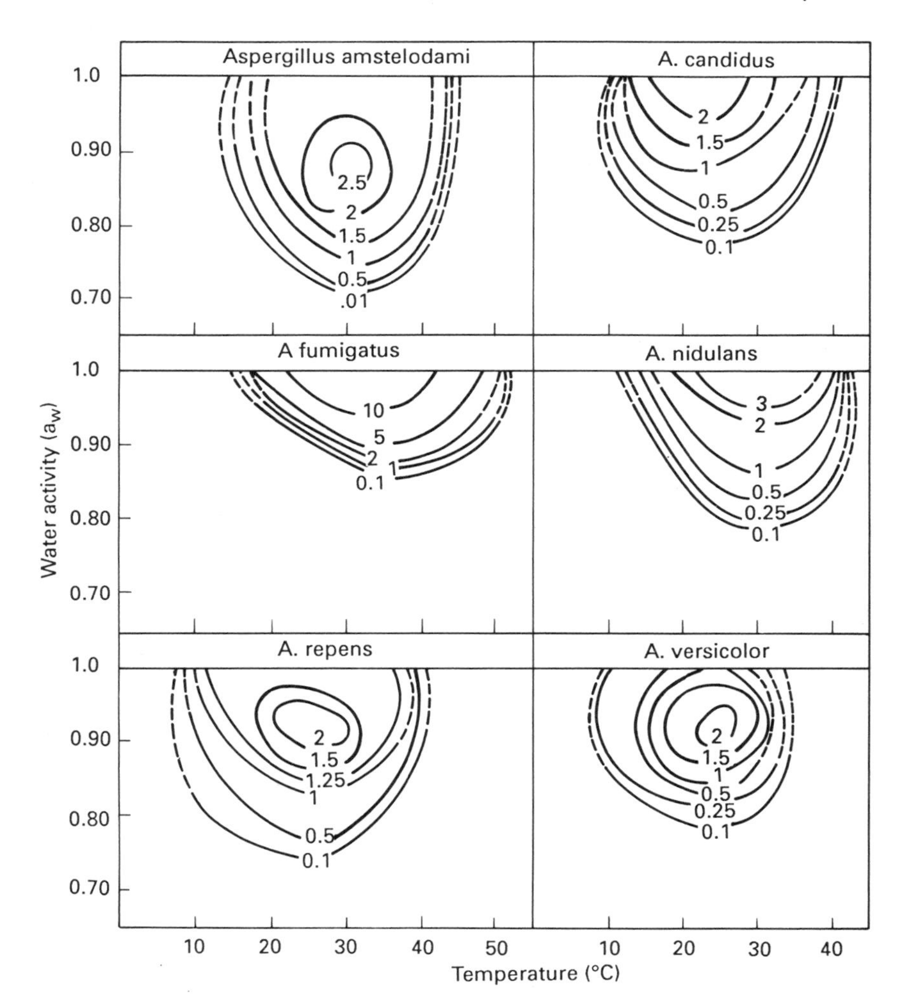

Fig. 5.2 Effect of temperature and water activity on the growth rate of *Aspergillus* species. The numbers on the isopeths are growth rates in mm day^{-1}. (From ref. 58.)

Accepting the above caveats, it is possible, on the basis of laboratory studies, to classify microorganisms with respect to the effect of medium water potential on growth. A classification is given in Table 5.3. The reader should note that there are many generalizations in that classification. Thus it should not be assumed necessarily that all species of a genus nor all members of a group grow equally well subject to the same water potential. The generalizations have been made on the basis of a range of observations using media whose composition can be different and whose water potential has not always been manipulated in the same manner. It will be realized from studying the list that there is much emphasis on organisms whose activities can have significant financial consequences, i.e. wood decay fungi, plant pathogens and food spoilage microorganisms. Also the reader should note that virtually all the information is for growth at laboratory temperatures and at pH values, which may not necessarily relate to those of the environments in which the organisms are found. Figure 5.2 shows how growth rate in relation to water activity of a range of *Aspergillus* species isolated from soil can vary in relation to temperature.

WHAT IS AN OSMOPHILIC MICROORGANISM?

Christian[16] defined osmophilic yeasts as those capable of multiplication in concentrated syrups of a_w below 0.85 (a water potential of −22.4 MPa), whereas Pitt[64] defined xerophilic fungi as those 'capable of growth under at least one set of environmental conditions at an a_w below 0.85'. There are other definitions of a similar kind.[76] Here, a conscious decision has been made not to use a tight definition. This is because, as yet, we cannot be at all certain that there is a group of microorganisms which grow at low water potentials and which are physiologically and biochemically discrete from those which grow at higher water potentials. As will be seen, the choice of a water potential of −22.4 MPa would rule out of consideration here very many microorganisms, including virtually all bacteria, found in environments which would be considered by many biologists to have a relatively low water potential. Consideration of the functioning of this wider group of microorganisms, rather than the much smaller group as defined by Christian[16] or Pitt,[64] provides both valuable insights as to how microorganisms respond physiologically and biochemically to osmotic stress and important insights how the most osmophilic species grow at very low water potentials.

Finally, it should be noted that most osmophilic microorganisms can grow reasonably well at water potentials relatively close to zero. However, a few of the most osmophilic microorganisms, the xerophilic moulds *sensu* Pitt,[64] are most unusual in that as well as being able to grow at very low water potentials, they have often very low growth rates in media with water potentials close to zero.[4,65] Optimum growth is usually obtained in media of water potential in the range −7.0–14.5 MPa (Fig. 5.3).

Physiology of osmotolerance

GROWTH IN A MEDIUM OF CONSTANT WATER POTENTIAL

Determination of protoplasmic solute potential

Two key questions must be answered when considering the growth of a microorganism at low water potentials. First, what solutes contribute to the solute potential and second, if we are concerned with an eukaryote, what is the extent to which solutes are either distributed uniformly throughout the protoplast or compartmented? Compartmentation will be important with ionic solutes, high concentrations of which could have an effect on metabolism.

To answer the first question, one must analyse a suitable amount of the

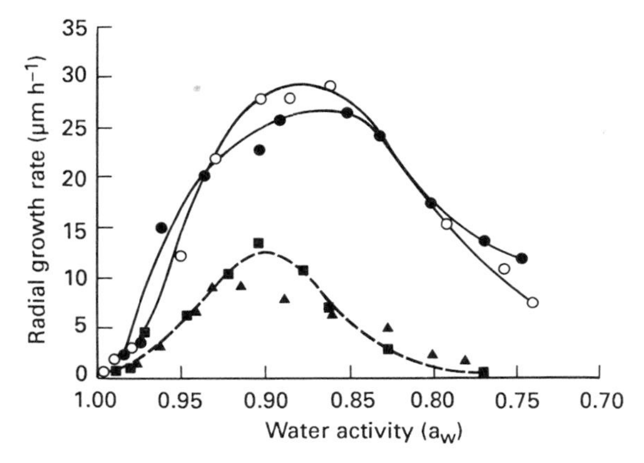

Fig. 5.3 Effect of water activity and solute on the radial growth of the xerophilic fungus *Basipetospora halophila*. Two isolates were studied, one represented by solid symbols, the other by open symbols. ○, ●, sodium chloride; ■, equal weights of glucose and fructose; ▲, glycerol. (From ref. 4.)

organism, and, on the basis of a knowledge of the water content of the protoplasm, determine the concentration of individual solutes within it. The solute potential obtained by totalling the potentials due to individual solutes should be equal to the potential measured independently, e.g. with a thermocouple psychrometer using frozen and thawed cells.[14,71] Further, in a wall-less cell, the solute potential should be equal to the medium water potential, whereas for walled cells, there should be a sufficient turgor potential for growth.

These determinations, though seemingly straightforward, can present difficulties. Eamus and Jennings[26] have discussed the problems with respect to filamentous fungi. Consideration of the difficulties as they relate to this group of microorganisms is appropriate, since difficulties of a similar, but not necessarily as extreme, kind exist for other groups. First there is the problem of determining the volume of water in the protoplasm. It should be noted that a value for the protoplasmic water content cannot be determined by subtracting the dry weight from the fresh weight. Water adhering to and in the wall will make a significant contribution to the total water content and must be determined by independent means. The best way is to use a solute which can enter all the water-filled spaces of the wall, yet not move across the plasmalemma. The procedure involves taking mycelial pellets with all but surface medium in which they have been grown removed and immersing them in a solution of a compound whose concentration can be determined readily. The pellet with surface liquid is then placed in water of known volume and the concentration determined of the compound released from the pellet. Knowing the original concentration, the amount of water in the wall can be readily calculated. At present, there is no compound which one can be certain absolutely fulfils the criteria of becoming associated solely with the wall water and having plasmalemma impermeability. However, blue dextran and inulin seem to be appropriate to use, the concentration of the former being determined spectrophotometrically and that of the latter by radioactive assay using a ^{14}C-labelled compound. The seeming appropriateness of these compounds needs to be assessed critically by independent procedures for estimating water in the wall. Even without that assessment, the procedure for determination of the water content of the protoplasm is critically dependent on the determination of the fresh weight of the mycelium.

There are considerable problems in determining the fresh weight of mould mycelium. As much as possible of the external medium has to be removed without damaging the mycelium. Any washing of the mycelium must be done with great care, particularly if the medium has a low solute potential. Hyphae growing in such media might burst when the external solute potential is brought to zero.[75] It is preferable that any washing is carried out at 0 °C.[5] Whether one is removing medium or water from mycelium prior to determining the fresh weight, the best procedure for removal is by gentle suction. This is rapid and reproducible though, as with determination of the wall water, there has been no really critical examination of the procedure.

Finally, there is the matter of ions and other solutes in the wall. Removal of as many of these as possible is preferable, particularly if one wishes to know the extent to which a chosen solute, which might be within both the medium and the protoplasm, contributes to the solute potential of the protoplasm. Non-ionized compounds are for the most part readily removed by washing, unless there is some relatively specific binding site within the wall. But cations can be bound very strongly. If the medium water potential has been lowered by increasing the concentration of a salt, then the relevant cation will be at high concentration in the wall. The best method for determining that concentration is by charging up the mycelium with the radioactive ion at known specific activity and determining the amount of radioactivity in the wall from a study of the loss of the isotope over time into a medium of the same composition as the charging solution but not containing the isotope.[88] When the mycelium has been grown in the presence of a mixture of cations, a possible procedure is to wash the mycelium rapidly with a relatively high concentration of a bivalent cation not present in the growth medium, e.g. lanthanum. That cation should replace other cations in the wall such that the amount of any of those present in the mycelium must relate to the protoplasm.

In the case of unicells, the procedures are somewhat more simple. This is because they can be centrifuged, if necessary through oil to remove the bulk of the medium, and the volume and weight of the compacted cells determined. Water in the wall and entrapped by the cells can be determined as for mycelial fungi, as can the concentration of cations in the walls. With unicells it is possible to determine intracellular non-osmotic volume by using the Boyle-Van't Hoff relationship.[68] The cells are made non-turgid under which circumstances the osmotic volume is directly proportional to the reciprocal of the external solute potential (osmotic pressure). A plot of the former against the latter allows determination of the osmotic volume when the reciprocal of the external solute potential is zero.

Ideally it should be possible, knowing the concentrations of each of the individual solutes (or at least each of the major ones) within the protoplasm, to calculate the protoplasmic solute potential (Ψ_s) from the following

$$\Psi_s = -RTmn\Phi,$$

where R is the gas constant; T is the temperature (K), m is the molality; n is the number of ions per molecule; Φ is the osmotic coefficient.[32] For a walled cell which is growing, there is a guideline to indicate if all the major solutes have been taken into account. Under such conditions, the calculated protoplasmic Ψ_s should be less than the water potential of the medium, i.e. there should be a positive Ψ_p which will mean that there is the necessary turgor (hydrostatic) pressure for growth. Of course, Ψ_s so calculated should agree with that determined by thermocouple psychrometry.[26]

Solute potential of microorganisms

In spite of the very considerable interest in osmophiles, we have surprisingly little data on how their total internal solute potential is generated in media of low water potential. Even then, the data relate only to fungi. Table 5.4 presents information about the percentage contribution of the major classes of solutes to the protoplasmic solute potential either calculated from the concentrations of the major solutes or determined directly. One should note that the values for the contribution of individual solutes are for those selected for analysis; in most cases they do not represent the full complement of solutes within the protoplasm and therefore the values given are likely to be an underestimate of the protoplasmic solute potential. In the case of ions, it is for *Aspergillus nidulans* and *Dendryphiella salina* that some account has been taken of cations which may be bound to the wall. In the case of the former fungus, cations were displaced from the wall prior to extraction of the mycelium by washing it with calcium chloride solution isoosmotic to the growth medium. In the case of the latter fungus, sodium in the wall of mycelium growing in the presence of −1.92 MPa sodium chloride was estimated using ^{24}Na.

With respect to the information given, that for *A. nidulans*, *D. salina* and *Thraustochytrium aureum* is the most realistic. This is because, in the case of the first two fungi, the calculated protoplasmic solute potentials are less than the medium water potential, thus turgor is being generated. In the case of *T. aureum*, the protoplasmic solute potential is very similar to the medium water potential. This would be anticipated since the cell (=sporangia) of this lower fungus is essentially wall-less. There is a word of caution. Any calculation of solute potentials from the concentration of individual solutes assumes that each one is equally active osmotically, i.e. that the outer membrane of the cell is essentially impermeable (relative to water) to each solute. In biophysical terms, we are assuming that the reflection coefficient of the membrane for all solutes under consideration is unity (it is zero when the membrane is equally permeable to a solute and to water).[62]

In the light of the above, there must be doubt about the data for *Chrysosporium fastidium* and *Penicillium chrysogenum* grown in −10 MPa glucose, since the calculated protoplasmic solute potential is *greater* than both the measured

Table 5.4 Percentage contribution, where known, to the solute potential (Ψ_s) of major classes of solute within the protoplasm of various fungi together with values for the osmotic potential either calculated from the concentrations of the known solutes or determined directly from the mycelium

Organism osmoticum added to basal medium (water potential)	*Contribution as*					*Identity of major organic solute*	Ψ_s		*Reference*
	Carbo-hydrates	*α-amino nitrogen*	*Inorganic ions*	*Organic acid*	*Major organic*		*Calculated (−MPa)*	*Measured (−MPa)*	
Aspergillus nidulans[a] NaCl(−3.86 MPa)	67	—	32	—	26	mannitol	4.63		6
Chrysosporium fastidium[b] Glucose(−10 MPa)	53	—	11	—	24.5	glycerol	7.6	11.85	55
Dendryphiella salina[a]									
NaCl(−0.96 MPa)	28	31	41	0	14	glycerol	1.63	—	88
(−1.92 MPa)	27	14	59	0	14	glycerol	1.63	—	
$MgCl_2$(−0.96 MPa)	29	20	51	0	19	mannitol	1.88	—	
(−1.92 MPa)	19	11	66	4	8	mannitol	4.32	—	
Inositol(−0.96 MPa)	26	34	32	8	14	inositol	1.58	—	
(−1.92 MPa)	33	26	33	8	23	inositol	2.30	—	
Penicillium chrysogenum[b]									
Glucose(−10 MPa)	44	—	15	—	24	glycerol	7.8	13.4	55
KCl(−10 MPa)	20	—	74	—	14	glycerol	11.9	12.7	
Phytophthora cinnamomi[b]									
Sucrose(−2 MPa)	33	32	58	—	33	sucrose	3.9	3.15	56
KCl(−2 MPa)	0	35	187	—	35	proline	7.7	3.43	
Thraustochytrium aureum[a]									
Seawater 50%(−1.24 MPa)	1	9	90	0	0.5	proline	1.36	—	87
75%(−1.85 MPa)	1	11	88	0	0.5	proline	1.78	—	

[a]Percentage of calculated
[b]Percentage of measured
— Not determined

Table 5.5 Concentrations of certain solutes within microorganisms which have been growing in medium of reduced water potential

Organism	*Solute*	*Concentration in organism (M)*	*Major solute in external medium and concentration*		*Medium (−MPa)*	*Reference*
Prokaryotes						
Klebsiella pneumoniae	glycine betaine	0.6	NaCl	800 mM		49
Synechocystis DUN 52	glucosyl glycerol	0.02	seawater	200‰		59
	quaternary N compounds	2.98				
	K^+	0.21				
	Na^+	0.23				
Fungi						
Debaryomyces hansenii	glycerol	1.04	NaCl	855 mM		68
Saccharomyces cerevisiae	glycerol	1.21				
Zygosaccharomyces rouxii	glycerol	1.19				
Wallemia sebi	glycerol	1.09[a]	NaCl		30.7	38
	sorbitol	1.94[a]	sorbitol		30.7	
Xeromyces bisporus	sorbitol	2.25[a]	sorbitol		30.7	

[a]Calculated from a fresh weight: dry weight ratio of 2.5

potential and the medium water potential. On the other hand, it seems likely in the case of *Phytophthora cinnamomi* that there are considerable errors in the determination of the concentrations of some of the solutes since the so-calculated solute potential is so much less than the water potential of the medium. The most likely possibility is that a considerable amount of potassium is bound to the wall.[56]

Table 5.5 provides information about the concentration of selected solutes in other microorganisms which in this case includes two prokaryotes, the bacterium *Klebsiella pneumoniae* and the cyanobacterium *Synechocystis* DUN 52. The fungi listed are amongst those which are the most xerotolerant.

If we take the data as a whole, the following general points can be made. Both ions and organic solutes can make a significant contribution to the protoplasmic solute potential. The extent to which either class of solute plays a major role depends on the organism. Thus, in *D. salina* and *T. aureum*, ions make the bigger contribution to the solute potential. On the other hand, in *A. nidulans*, organic solutes, in particular carbohydrates, make the bigger contribution. When ions are used, one is dealing with solutes which are absorbed from the external medium in order to help generate the requisite solute potential. Organic solutes present in the medium can also be used in this manner, *vide* the considerable contribution of sorbitol to the protoplasmic solute potential of *Walemia sebi* and *Xeromyces bisporus* grown in a medium with a water potential of −30.7 MPa generated with the same compound. The internal concentration of sorbitol in the case of the former fungus after around 60 days growth was over five times that of the solute next highest in concentration, namely glycerol, whereas with the latter fungus there was six times as much sorbitol as glycerol.[39] Likewise, there was as much glucose as glycerol in *C. fastidium* grown in a medium with a water potential of −10 MPa generated with glucose, whereas in *P. chrysogenum*, the glucose concentration was half that of glycerol in a similar medium. Yet when the latter fungus was grown in a medium of −10 MPa generated with KCl, no glucose was present in the mycelium.[55] A similar situation to *P. chrysogenum* and glucose held for *Phytophthora cinnamomi* with respect to sucrose when that solute was used to reduce the medium water potential.[56] Finally, it has been shown that when the osmophilic yeast, *Debaryomyces hansenii*, which normally accumulates glycerol when grown under saline conditions with glucose as the carbon source, produced no glycerol when *meso*-erythritol replaced glucose, the polyol becoming the major organic solute.[60]

Studies on enteric bacteria

It is appropriate here to refer to recent studies on the response of two enteric bacteria *Escherichia coli* and *Salmonella typhimurium* to osmotic stress.[35] Under normal conditions, potassium is the primary osmotic solute but the internal concentration of the ion decreases with decreasing external water potential. In *E. coli* at least two separate transport systems have been shown to be responsible for potassium influx.[8,28,69] There is a constitutive system of low affinity (Trk) and an inducible system of high affinity (Kdp). More knowledge has been accumulated about the latter system which has been shown to be ATP-driven and derepressed in low potassium media. The protein involved belongs to the family of ATPases involved in cation transport. The Trk system has a very low affinity for potassium (a K_m of 1.5 mM as opposed to 2 μM for Kdp) but a high rate of transport such that cellular potassium can be rapidly replenished. Evidence to date indicates that active accumulation of potassium by Trk is achieved through the proton motive force. ATP is required but in a regulatory capacity. Genetic evidence points to at least three separate systems involved in potassium exit from the cell. Two of the mutant classes are affected with respect to potassium efflux *per se* rather than with respect to net potassium uptake.

High turgor inhibits the influx systems and activates at least one of the efflux systems; though it must not be concluded that turgor pressure *per se* is the signal (see later). Above 0.25 osM, other solutes play an increasing role in the generation of the internal solute potential of the cells. In particular, there are trehaloses which the cells can synthesize and betaine which the cells of many strains cannot synthesize but can accumulate from the medium. There is a high affinity betaine transport system which is expressed only in the presence of high internal concentrations of potassium.[74] Thus betaine is scavenged from a medium of low water potential and accumulated in the cells such that turgor is maintained.

It is not appropriate to go further into this relatively detailed picture of how turgor is regulated in *E. coli* and *S. typhimurium*. Here the point to be emphasized is that a microbe can respond to a decreased medium water potential by absorbing solutes from the medium specifically for that purpose. The studies on the two enteric bacteria indicate that special transport systems may be switched on when there is a low medium water potential. It is clear that those examples to which I have referred in Tables 5.4 and 5.5 in which solutes are absorbed unchanged to generate the required solute potential in the face of a low medium water potential need to be investigated further. As with the enteric bacteria, solute accumulating systems may be induced in media of low water potential. There is some evidence that this is so when *Debaryomyces hansenii* is grown in highly saline media. Cells which have been grown in such media possess an ATPase having different properties (such as stimulation by sodium ions) from and additional to the normal proton-translating ATPase found to date in fungi.[19]

COMPATIBLE SOLUTES

The frequent presence of high concentrations of non-ionic organic solutes has been interpreted in terms of the ability of many of the compounds, even when at relatively high concentration, not only to make an important contribution to the protoplasmic solute potential but at the same time have minimal effect on the conformation of proteins. Compounds which have this property have been termed *compatible solutes*.[13] *In vitro* studies on enzymes have shown that sometimes they are protected to a degree by these compounds from the deleterious effects of salts, particularly NaCl. The latter property may be termed *haloprotection*.[85] Of the compatible solutes, it appears that only glycine betaine can act as an effective haloprotectant.[85] It needs to be noted that in many cases it is not clear from those studies which have been made how far the concentration of compatible solute used relates to its protoplasmic concentration. In one instance, where results are relevant to the living microorganism, namely the marine filamentous fungus

Table 5.6 Details of compounds, thought to act as compatible solutes, produced by microorganisms under stress. Information obtained mainly from refs 38, 40, 42, 67, 81, 87, 88

Compound	*Species or class of microorganism*
Glycerol	*Debaryomyces hansenii* (osmophilic yeast) *Dendryphiella salina* (marine mould) *Zygosaccharomyces rouxii* (osmophilic yeast) Xerophilic fungi
Arabitol	*D. hansenii* *D. salina*
Erythritol Mannitol	*D. salina*
Glycosyl glycerol	Marine cyanobacteria
Sucrose	Freshwater cyanobacteria
Trehalose	Freshwater cyanobacteria *Escherichia coli*
Proline	Lower fungi, including the marine fungus *Thraustochytrium*
Glutamate	Bacteria
Glycine betaine	Hypersaline cyanobacteria
Proline betaine	*E. coli* in the presence of choline
Glutamate betaine	Hypersaline cyanobacteria
Ectoine	*Ectothiorhodospira halochloris*

Dendryphiella salina, it has been found that the compatible solute glycerol, at its protoplasmic concentration, did not protect malic dehydrogenase from the deleterious effects of sodium chloride at high concentrations in the assay medium.[63]

Table 5.6 lists those compounds produced within microorganisms which are thought to act as compatible solutes. By far the best consideration of how macromolecular structure and function is maintained in the face of low solute potential brought about by these solutes is that of Yancey *et al.*[93] The understanding of how these solutes may act is enhanced by knowledge of how inorganic ions may exert their effects on proteins in solution and, for relevant information, the reader should consult refs 85 and 91. Indeed, valuable comparisons have been made between the similarity of certain compatible solutes and those ions which have been shown to favour the native states of proteins.

Although the concept of the compatible solute is a valuable one, it is probable that it is simplistic to consider these solutes as having solely an osmotic role. Thus polyols, whose physiology and metabolism in fungi and green plants had been studied in much detail prior to any consideration of them having a role as compatible solutes, were thought to serve the following roles: (i) act as carbohydrate reserves; (ii) act as translocatory compounds; (iii) play a role in coenzyme regulation and storage as well as helping to generate turgor.[53] With increasing knowledge of the biochemistry and physiology of polyols in fungi, it is becoming clear that these compounds are part of the machinery used by these organisms to maintain a constant physiological and biochemical milieu in the cytoplasm. It is for this reason, Jennings[43] has referred to polyols as *physiological buffering agents*. Nevertheless, that concept should not deter one from accepting the much simpler view that a compound such as glycerol has a major osmotic role as has been found to be the case in the osmotolerant yeasts *Zygosaccharomyces rouxii*[27] and *Debaryomyces hansenii*.[1] For both organisms, glycerol content has been shown to be close to proportional to sodium concentration in the growth medium.

RESPONSE OF A MICROORGANISM TO A CHANGE IN MEDIUM WATER POTENTIAL

Introduction

The previous section has been concerned with growth of microorganisms in media of constant water potential. While such conditions are not strictly steady-state, the time-scale of change is such that they can be accepted as such. Under natural conditions, there might be rapid changes in water potential so that the microorganism is no longer in equilibrium with its environment. If the water potential decreases, water will tend to be drawn out of the organism; conversely, if the water potential rises, water will tend to enter. In either case, unless there is some procedure for adjustment of the internal water potential, the microorganism cannot continue to function normally. Of the two situations, entry of water can result in the starker consequence, namely bursting of the cell. Reference has already been made to the bursting of hyphal tips, when the mycelium of filamentous fungi is exposed to a higher external water potential. However, in this particular instance, the consequences need not be irremedial, since in septate hyphae the sub-apical or older compartments of the hyphae can be sealed off from the bursting apical compartment by blockage of the septal pore by a Woronin body, a proteinaceous inclusion within the cytoplasm.[26] Under such circumstances, renewed growth is achieved by older compartments producing new branch hyphae.[55,79] Where water is drawn out of the cell, there will occur membrane deformation and concentration of solutes within the cytoplasm. Under such circumstances, the cell is being dehydrated and this is a matter that will be returned to.

Terminology

When considering adjustment of a microorganism to a new medium water potential, there is a need to realize that there can be two points of reference for that adjustment. The cell can regulate to a set volume or it can regulate to a set turgor potential or pressure. The matter has been well discussed with particular reference to green plant cells.[21,66,94] In these cells, it has been found frequently that, when the water potential of the medium is changed, the solute potential changes such that the turgor potential is restored close to a value achieved in the original medium. On the other hand, in 'wall-less' cells, there is evidence that it is volume which is restored to the initial value. This has led to the use of two terms *turgor regulation* and *volume regulation.*[21] That there are these two terms assumes that there is no relation between turgor and volume. But that is not so. Changes in both turgor and volume depend on the elastic properties of the plasmalemma and cell wall. Thus it is only in wall-less cells that we can speak with some confidence about volume regulation. In walled cells it is not clear whether it is turgor or volume which is the output of the control process. Further, increasing knowledge of how green plants, particularly the algae, respond to changes in medium water potential have shown that there can be incomplete restoration of both turgor and volume (depending on the organism) following a change in medium water potential.

This means, as Reed[66] has pointed out, that it is not easy to produce terms which are necessarily appropriate to all situations. He has suggested that qualifying adjectives can be used. Thus one can speak of *full* or *complete or incomplete volume regulation.* Equally *turgor maintenance* might be used if positive turgor is still generated subsequent to a change in medium water potential. However, it is very clear that *osmoregulation* (or *osmotic regulation*) a term much used by plant physiologists and microbiologists alike is not appropriate unless it can be unequivocally demonstrated that osmotic potential is the reference point. Reed favours the more general use of the term *osmo-acclimation*, using other terms like *turgor regulation* (*full* or *partial*) and *volume regulation* (*complete* or *incomplete*) only when the data are sufficient to indicate that such regulation is taking place.

Two other terms need to be defined at this juncture. They are *upshock* and *downshock*. Both refer to changes in the *concentration* of osmotica in the external medium. *Upshock* (or *hyperosmotic* shock) indicates an *increase* in concentration and therefore a *decrease* in water potential; *downshock* (or *hypoosmotic* shock) refers to the converse.

Experimental studies

We know very little about how microorganisms adjust to fluctuations of external water potential. Yet there is now a significant literature on the response of algae to changes in medium water potential, particularly with reference to those species which are intertidal and estuarine, and which therefore must be capable of osmoacclimation.[7,34] Microbiologists need to be aware of this literature, because it provides indications as to possible processes, in particular fluxes of ions into and out of cells, which allow such organisms to osmoacclimate. In a similar way, the literature about the guard cells of stomata which are the prime example of cells which routinely undergo very considerable osmotic changes, also provides powerful insights. In these cells, there are very large changes in volume and turgor as well as large changes in salt content. Therefore, of necessity, there are marked changes in the fluxes of individual ions into and out of the cells.[57]

Turning to microorganisms, the few experimental studies have concerned fungi. Luard,[54,56] by transferring mycelium on cellophane from one agar medium to another whose water potential differed by the presence or absence of appropriate concentrations of glucose, studied the effect of upshock and downshock on *Chrysosporium fastidium* and *Penicillium chrysosporium*. Both upshock and downshock brought about cessation of growth. In the latter case, there was bursting of hyphal tips and, if growth recurred, it was some distance (at least 250 μm) behind the apex. In the case of upshock, new growth, provided that the magnitude of the shock was not greater than −10 MPa, occurred first with branching of the apex.

Where mycelial solute changes were determined, it is interesting to see that, when an upshock of −10 MPa was given to *P. chrysogenum*, there was very little change in the content of potassium or sodium (though the concentration probably increased very significantly due to water loss from the mycelium). On the other hand, there was a very considerable synthesis of glycerol; within 8 h, the content of the mycelium was 53% of that found in mycelium growing *ab initio* on medium to which glucose had been added to increase the water potential by −10 MPa. With downshock, for both fungi, there was a fall in the content of organic solutes. Glycerol content showed the biggest magnitude of change, with 41% decrease in the case of *P. chrysogenum* but 87% in the case of *C. fastidium* over 4 h. In both fungi, there was a rapid fall in the glucose content with little change. It is believed that in *P. chrysogenum* fall in glycerol content was due to metabolism, whereas the fall in glycerol and glucose in mycelium of *C. fastidium* was due to outflow across the plasma membrane into the medium.[56] Some of the fall must be due to loss as the result of bursting of hyphal tips but it does seem likely that within unruptured membranes some permeability changes occurred in hyphae of *C. fastidium* during downshock, since over the first 30 min the K : Na ratio changed from 8.3 to 1.3.

It seems likely that release of glycerol into medium is an important element of response of an osmotolerant fungus to an osmotic downshock. It has been shown that when cells of *Debaryomyces hansenii* were grown at high salinity there was a rapid loss (less than 10 min) of glycerol from the cells, the amount lost being increasingly greater the more dilute the medium into which the cells were transferred.[1] At this stage, there is no indication as to how loss might be controlled. When cells of *D. hansenii* were exposed to an upshock, glycerol accumulated in the cells. The levels reached were in direct proportion to the level of salinity used to bring about upshock and the amount produced was sufficient to generate the appropriate internal solute potential. When other solutes were tested, similar levels of glycerol were produced in the cells for any upshock leading to a particular water potential. A proportion of the glycerol produced entered the medium. Brown[10,12] has reported that the osmophilic yeast *S. rouxii* when grown in saline media retains more glycerol, although not necessarily producing more in total, than the more osmosensitive *Saccharomyces cerevisiae*. So one presumes the ability of *D. hansenii* to withstand upshock will be a function not only of its ability to produce glycerol but also to retain the compound.

The time taken for the final level of glycerol to be achieved was found to be around 2–4 h, about a generation time. It is probable from other studies on *D. hansenii*[37] that the initial reaction to upshock is a rapid net influx of sodium chloride into the cell. This is followed by the slower exchange of potassium for sodium. But these changes need examining in more detail. It is sufficient here to point out that in *D. hansenii*, the production of glycerol is the means by which the cells regain a steady state osmotically, when they are exposed to upshock. The short-term response aimed at maintaining turgor seems to be via movement of ions into the cell. On the other hand, it is clear that, when the cells are exposed to a downshock, glycerol is involved in the short-term response, being lost into the medium.

The probable importance of monovalent cations in turgor regulation seems likely to parallel what has been found for *Escherichia coli*.[8] In this bacterium, the extent of potassium uptake is determined by the osmolarity of the growth medium. Sudden upshock increases potassium influx through stimulation of one or other of the two independent uptake mechanisms. Likewise, downshock stimulates efflux, through three possible systems. After upshock, the internal potassium concentration falls. The rate of fall is rapid if betaine is present in the medium (or choline which can be converted to betaine within the cell). Eventually, the potassium content can return to a value close to that in the cells before upshock. The betaine transport system is switched on by the high potassium concentration within the cell. Choline uptake appears to be similarly regulated.[73] In the absence of betaine or choline, the rate of fall of internal potassium is much slower due to the slow synthesis of trehalose.

Future studies

There should be no doubt, after consideration of the above, that there is a need for much more information about how osmophiles react to a change in the water potential of the medium. There is a clear need to establish how the solute content of the organism changes as the cells or mycelium move towards the new steady state. In the experiments of André *et al.*[3] it was clear that the glycerol was produced from exogenous glucose. It is possible that insufficient carbon reserves were present in the cell for there to be a sufficient flux of carbon from those reserves to glycerol when exogenous glucose was absent. It would be interesting to find out whether cells which had been grown in a manner devised to produce a high internal content of reserve carbon would behave differently. There is certainly evidence from other organisms of the conversion of carbon reserves into osmotically active solutes, the most notable being the golden-brown alga *Poterioochromonas malhamensis*.[45,46] This alga, when given an upshock, initially shrinks (over 1–2 min) due to water loss but regains volume after 1–2 h due to water inflow caused by the production of the solute isofluoridoside (α-galactosyl-1,1-glycerol) which can be formed by photosynthesis or from exogenous glucose or from the reserve β-(1→3)-glucan (chrysolaminarin). With decrease in the medium water potential, all the carbon is transferred into chrysolaminarin thus allowing the cell to readapt by decreasing its solute potential. What is so interesting about this biochemical response to osmotic change in the medium is that it is signalled by calcium as a consequence of alteration in the calcium pools in the cell consequent upon change in cell volume. Incidentally, in *P. malhamensis*, when exposed to an upshock with sucrose, there is a rapid net influx of potassium indicating once again the importance of ions in the short-term readjustment of cells to changes in medium water potential.[47]

This matter of the interrelationship between insoluble carbon reserves and soluble osmotically active carbon compounds needs to be investigated much further in osmophilic microorganisms, both prokaryote and eukaryote. With respect to the former, we know that in a moderately osmophilic cyanobacterium *Spiralina platensis* upshock leads to the conversion of glycogen to glucosyl–glycerol while the converse occurs in downshock.[84]

Commercial aspects

Osmophiles are of great economic importance as food spoilage organisms. All foods contain water and those foods most likely to rapidly deteriorate due to both chemical and biological change are usually those with a high water content. Food can be preserved for considerable periods by freezing. But this is not always possible, due either to the deleterious effects of freezing or for economic reasons. In any case, the more traditional means, curing or salting or syruping or sugaring, designed to reduce the available water by lowering the water activity or water potential, produce foods having distinctive flavours which are esteemed in their own right. Even though these means are used to prevent spoilage, it can nevertheless occur. Equally, dehydration can be a way of preserving food. However, there are times when the moisture content of dehydrated food becomes accidentally raised. For all foods preserved or dehydrated, there is now much information about possible microbial contaminants in relation to water activity.[17,24,25,76,77,80] The reader is advised to turn to the relevant literature for detailed information about the microorganisms involved and to obtain the proper background to modern procedures for reducing water activity in food.

Nevertheless, there are some matters which need to be highlighted here. Although it is seemingly relatively easy to determine the growth characteristics of a microbe in relation to medium water potential, interpretation of the results in relation to the food in question is not necessarily easy. As has been indicated above, the ability to grow in a medium of low water potential can depend on the solutes available. It is not always clear in particular samples of food what solutes might be available. In any case, matric forces can be an important component of solid foods. Indeed, it may well be the availability of solutes which allow yeasts to grow in liquids with water activities between 0.75 and 0.65 but not on solid material of the same water activity. In the latter, absorbable solutes are likely to be less readily available and the matric potential very significant.

There is a need to remember that the contaminating organism can change the solute characteristics of the food by enzyme activity. It is possible that such an organism might change the water potential of the food. It is well known that osmophilic moulds are more likely to cause spoilage of solid foods than osmophilic yeasts.[76] No satisfactory explanation has been given for this difference. One possibility is that a filamentous fungus is able to initiate growth at a particular point, where there is a favourable water potential, then translocate water through its hyphae to raise the water potential of the medium some distance away. That this is possible has been demonstrated for the wood decay fungus *Serpula lacrimans* and is likely to be of relatively general occurrence as indicated by the widespread ability in fungal colonies of aerial hyphae to produce droplets, the liquid for which can only have come via the hyphae from the medium.[44]

Finally, when considering the water potential of a particular food, there are problems if it is in any way heterogeneous. Conceivably, depending upon the mobility of water, there may be differences in water potential throughout the material. Equally, there is the possibility that the extent to which solute or matric potentials make up the water potential will differ from one part of the material to another.

Genetic aspects

This is an area of study which is just being opened up. At present, the only truly genetical information of relevance relates to moderately tolerant prokaryotes. Reference has been made earlier to the expression in the two enteric bacteria, *Escherichia coli* and *Salmonella typhimurium*, of a high affinity betaine transport system which is expressed when there is high material potassium. It has been shown that in *S. typhimurium* the *proU* locus, for the osmotically induced betaine transport, encodes a 31 kDa periplasmic protein whose synthesis is induced by osmotic stress.[36] A specific betaine-binding activity with a K_D of about 1 μM was found to be present in the periplasm of induced cells, which was absent from those *proU* mutants lacking the 31 kDa periplasmic protein. Thus osmotic stress leads to the production of a specific transport protein.

Very interesting studies of a related kind concern the ability of exogenous choline and glycine betaine aldehyde to stimulate growth of *E. coli* under osmotic stress almost as effectively as glycine betaine; choline has no osmoprotective effect.[73] Under aerobic conditions choline is oxidized to glycine betaine aldehyde which in turn (under aerobic or anaerobic conditions) is oxidized to glycine betaine. The activities of the two enzymes, choline dehydrogenase and glycine betaine aldehyde dehydrogenase respectively are both stimulated by increased concentrations of sodium chloride in the external medium.

When cells are transferred from a medium of low salinity to one of high salinity, expression of increased enzyme activity is inhibited by chloramphenicol, indicating *de novo* synthesis of enzyme. Similarly, there is induction of two independent choline transport systems (one high and one low affinity) with salinity, their induction being inhibited also by chloramphenicol. Genetic studies indicate that the genes for the dehydrogenases and the high affinity uptake system are clustered on the chromosome but appear to represent different transcriptional units. While all require increased external salinity, only the genes for the dehydrogenases require choline as an inducer.

Proline is synthesized in *E. coli* and *S. typhimurium* from glutamate via the coordinated function of three enzymes: γ-glutamyl kinase (product of the *proB* gene), glutamate-γ-semialdehyde (product of the *proA* gene) and pyrroline-5-carboxylate reductase (product of the *proC* gene). Using a *proBA* deletion mutant of *S. typhimurium* which harboured *proBA* genes of *E. coli* on an F′ plasmid, mutants have been isolated which were resistant to the toxic analogue of proline, L-azetidine-2-carboxylate.[22] One mutant which was isolated was found to accumulate 32 nmol proline g protein^{-1} as against less than 1.2 in the control strain. Under saline conditions, the level in the mutant rose to 785 nmol g protein^{-1} with a barely detectable increase in the control. Within the mutant, γ-glutamyl kinase has a reduced sensitivity to feedback inhibition by proline which was independent of the source of glutamate–semialdehyde dehydrogenase present.[72] Nevertheless, the ability of the latter enzyme in the mutant to interact with the kinase was altered in thermal stability, though how this alienation relates to osmotolerance is not clear.

The proline overproducing mutation was located in the *E. coli* episome F'_{128}

harboured by *S. typhimurium*. Conjugal transfer of the mutant F′ plasmid to different strains of enteric bacteria such as *Salmonella* sp. and *Klebsiella pneumoniae* resulted in proline overproduction and increased osmotolerance.[22,52] Later studies have led to the construction of a recombinant plasmid conferring proline overproduction and increased osmotolerance.[41]

It is clear that it is now possible to engineer genetically osmotolerant bacteria and there is a view that *Rhizobium*, which is very osmosensitive, is a good candidate for such engineering.[50,51] If *Rhizobium* could be so engineered, it is argued that it should then be possible to produce legumes better able to grow in more saline soil. But the legumes themselves must be made more salt-tolerant; these crop plants are more than usually salt-sensitive.[29] Further, we need to be careful about what may happen to such genetically engineered microbes in the environment. It is possible that bacteria unfavourable to legume growth or indeed pathogenic fungi might acquire genetic material for osmotolerance.

For eukaryotes, knowledge of the genetics of osmotolerance is virtually nil. Nevertheless, mutants are now being studied. Mutants have now been produced which can grow on glucose but not on glycerol and their ability to grow in glucose media of increased salinity has been studied.[1] When subjected to such increased salinity, the internal glycerol content increased linearly in proportion to salinity as with the wild-type strain, as indicated earlier. However, as the salinity was increased, the growth rate became much lower than that of the wild type. More glycerol was lost in the medium from the mutant. Uptake of glycerol was not affected but the K_m of *sn*-glycerol 3-phosphate dehydrogenase in the mutant was increased 33-fold for *sn*-glycerol 3-phosphate. Further studies in this mutant[3] have shown that at very high salinity (2.73 M NaCl) growth in 0.5% glucose medium is inhibited completely. Yet in the presence of glycerol at a concentration as low as 0.5 mM, the growth rate became not far short of that of the wild type, although the lag phase was much longer. Increasing the glycerol concentration shortened the lag phase. Further studies showed that the mutant requires a minimum concentration of glycerol to initiate growth at high salinities but thereafter the content of the cells is maintained by metabolic production. Compounds such as mannitol, arabitol, proline and betaine could not reproduce the effect of glycerol in causing growth to occur at high salinity. Interestingly, the presence of glycerol in the medium could also reduce the lag phase of wild-type cells in a medium of high salinity. The mechanism of action of glycerol is obscure. One might guess that the compound acts as an effector for some process which leads to the induction of mechanisms leading to an ability of the cell to handle the increased concentration of salt in the external medium. It is known that in *D. hansenii* there is a changed non-mitochondrial ATPase complement within the cells as a result of growth in saline conditions.[19] It is possible that such a change is triggered by the presence of glycerol in the cell.

There is a recent report on mutants isolated from *Zygosaccharomyces rouxii* which were less able to grow in saline media.[92] Most were to some extent less tolerant to high concentrations of glucose than the wild type. It appeared from studies of the mutants that the growth of yeast in media containing high concentrations of sodium chloride requires the ability to produce glycerol, to maintain the required concentrations in the cells, to take up glycerol which has been lost and to assimilate the compound. Much of the data presented are semiquantitative; it is to be hoped that selected mutants are subjected to more detailed study.

Future prospects

There is little doubt that we need much more information about osmophilic microorganisms. Anyone reading what has been written about them here will quickly see deficiencies in our knowledge, nowhere more so than for the xerophilic moulds. These fungi need much more study, particularly since we have no real idea as to why they are better able than other microorganisms to withstand the lowest water potentials. To date, much of our information about osmophilic microorganisms concerns their ability to produce compatible solutes. Future work must concentrate on establishing what other compounds contribute to the protoplasmic solute potential and how the various organisms respond to a change in the external water potential. The dynamics of response provide a powerful way of probing the underlying physiology of osmophily. Equally, there is a need to investigate growth under proper steady-state conditions in a chemostat. The availability of external carbon compounds for the production of compatible solutes and as a source of energy to maintain the necessary osmotic gradients across the outer membrane will have an important influence on the rate of growth which is best probed by continuous culture studies. Such studies are the most likely to be relevant to the environment in which a particular microorganism might live. However, although physiological studies are of high priority, continual study of the composition of osmophilic microorganisms is still important. Here there is a need to focus on the composition of membranes, since they must be composed in such a manner as to allow maintenance of structure in the face of highly dehydrating conditions. Finally, one trusts that the power of the genetic approach will continue to be exploited. There is an urgent need to increase our understanding of the genetic basis of osmophily in fungi.

Osmophilic microorganisms have a high profile economically in terms of their ability to spoil food. The question is how they might be used in a beneficial manner commercially. There is potential in their ability to produce at high concentration organic compounds which are distinct from secondary metabolites. But osmophilic microorganisms might have an important role in biotechnological processes in parts of the world where freshwater is at a premium. Ability to grow in saline and other solutions of high concentration, yet produce enzymes or secondary metabolites, could be of considerable commercial value.

Conclusions

In the Introduction, a number of names were given for those microorganisms which are able to grow under conditions in which water is not readily available. If it can be substantiated that there are microorganisms such as *Basipetospora halophila* that are only able to grow well at low water potentials, then such organisms correctly can be termed osmophiles. However, to date, there seem to be few such organisms. Most microbes which can grow in media of low water potential can also grow well in media with a water potential close to zero. One might argue that microorganisms of this kind be termed osmotolerant. If this were the case, it would seem wise to restrict the term osmophiles to those microorganisms which grow best in media of low water potential.

However, the above arguments are based on physiological considerations, from growth studies in the laboratory. If ecological criteria are used, then a much greater group of microorganisms could be categorized as osmophiles, i.e. microorganisms which can be isolated from and shown to be growing in environments of low water potential. Use of the term in this way would parallel the use of the term halophile for microorganisms isolated from very saline environments and halophytes for green plants which complete their life cycle in saline soils. In the strict sense, there is a great deal to be said for considering as osmophiles those microbes which can grow in environments with a water potential of −22.4 MPa (0.85 water activity). However, the information which has been brought forward above indicates that the use of −22.4 MPa as the water potential for categorizing those microorganisms that can be termed osmophilic must be considered as rather arbitrary. Though the information is as yet fragmentary, that which is available indicates that many osmophilic microorganisms differ from those which are categorized as non-osmophilic not in kind but in degree. Thus many of the physiological processes used to adjust the internal solute potential to a low medium water potential are of a similar kind in the two groups of microbe. It seems sensible, therefore, to think of osmophilic microorganisms as being a somewhat indeterminate group. This suggests that qualifying terms such as 'moderate' and 'highly' need to be used to give a little more precision to the term osmophile. However, irrespective of terminology, it must be clear from what has been presented in this chapter that there is a need for much more information about osmophiles. At present, that which we do have is all too fragmentary.

References

1. Adler, L., Blomberg, A. and Nilsson, A. (1985) *Journal of Bacteriology*, **162**, 300–6.
2. Anand, J.C. and Brown, A.D. (1968) *Journal of General Microbiology*, **52**, 205–12.
3. André, L. Nilsson, A. and Adler, L. (1988) *Journal of General Microbiology*, **134**, 669–77.
4. Andrews, S. and Pitt, J.I. (1987) *Journal of General Microbiology*, **133**, 233–8.
5. Bartnicki-Garcia, S. and Lippman, E. (1972) *Journal of General Microbiology*, **73**, 487–500.
6. Beever, R.E. and Laracy, E.P. (1986) *Journal of Bacteriology*, **168**, 1358–65.
7. Bisson, M.A. and Gutknecht, J. (1980) in W.J. Cram, R.M. Spanswick and J. Dainty (eds) *Membrane Transport Phenomena: Current Conceptual Issues*, Amsterdam, North Holland, pp. 131–42.
8. Booth, I.R. (1985) *Microbiological Reviews*, **49**, 359–78.
9. Briggs, G.E. (1967) *Movement of Water in Plants*, Oxford, Blackwell Scientific.
10. Brown, A.D. (1974) *Journal of Bacteriology*, **118**, 769–77.
11. Brown, A.D. (1976) *Bacteriological Reviews*, **40**, 803–46.

12. Brown, A.D. (1978) *Advances in Microbial Physiology*, **17**, 181–242.
13. Brown, A.D. and Simpson, J.R. (1972) *Journal of General Physiology*, **72**, 589–91.
14. Brown, R.W. and van Haveren, B.P. (1972) *Psychrometry in Water Relations Research*, Logan, Utah Agricultural Experimental Station.
15. Brownwell, K.H. and Schneider, R.W. (1985) *Phytopathology*, **75**, 53–7.
16. Christian, J.H.B. (1963) in J.M. Leitch and D.N. Rhodes (eds) *Recent Advances in Food Science*, Vol. 3, London, Butterworths, pp. 248–55.
17. Christian, J.H.B. (1980), in *Microbial Ecology of Foods, Vol. 1. Factors affecting Life and Death of Microorganisms, The International Commission for Microbial Specifications for Foods*, The International Commission for Microbial Specifications for Food (eds), London, Academic Press, pp. 70–91.
18. Clarke, R.W., Jennings, D.H. and Coggins, C.R. (1980) *Transactions of the British Mycological Society*, **75**, 271–80.
19. Comerford, J.G., Spencer-Phillips, P.T.H. and Jennings, D.H. (1985) *Transactions of the British Mycological Society*, **85**, 431–8.
20. Cook, R.J., Papendick, R.I. and Griffin, D.M. (1972) *Soil Science Society of America Proceedings*, **356**, 78–82.
21. Cram, W.J. (1976), in U. Lüttge and M.G. Pitman (eds) *Encyclopedia of Plant Physiology, New Series, Vol. 2. Transport in Plants II, Part A Cells*, New York, Springer-Verlag, pp. 284–316.
22. Csonka, L.N. (1981) *Molecular and General Genetics*, **182**, 82–6.
23. Dainty, J. (1969), in M.B. Wilkins (ed.) *Physiology of Plant Growth and Development*, London, McGraw-Hill, pp. 421–542.
24. Dallyn, H. and Fox, A. (1980), in G.N. Gould and J.E.L. Corry (eds) *Microbial Growth and Survival in Extremes of Environment*, The Society for Applied Microbiology Technical Series No. 15, London, Academic Press, pp. 121–39.
25. Duckworth, R. (1975) *Water Relations of Food*, New York, Academic Press.
26. Eamus, D. and Jennings, D.H. (1986), in P.G. Ayres and J.L. Boddy (eds) *Water, Fungi and Plants*, British Mycological Society Symposium 11, Cambridge University Press, pp. 27–48.
27. Edgley, M. and Brown, A.D. (1978) *Journal of General Microbiology*, **104**, 343–45.
28. Epstein, W. (1986) *FEMS Microbiology Reviews*, **39**, 73–8.
29. Greenway, H. and Munns, R. (1980) *Annual Review of Plant Physiology*, **31**, 149–90.
30. Griffin, D.M. (1972) *Ecology of Soil Fungi*, London, Chapman & Hall.
31. Griffin, D.M. (1977) *Annual Review of Phytopathology*, **15**, 319–29.
32. Griffin, D.M. (1981) *Advances in Microbial Ecology*, **5**, 91–136.
33. Griffin, D.M. and Luard, E.J. (1981), in M. Shilo (ed.) *Strategies of Microbial Life in Extreme Environments*, Weinheim, Verlag Chemie, pp. 49–63.
34. Hellebust, J.A. (1976) *Annual Review of Plant Physiology*, **27**, 485–505.
35. Higgins, C.F., Cairney, J., Stirling, D.A., Sutherland, L. and Booth, I.R. (1987) *Trends in Biochemical Sciences*, **12**, 339–44.
36. Higgins, C.F., Sutherland, L., Cairney, J. and Booth, I.R. (1987) *Journal of General Microbiology*, **133**, 305–10.
37. Hobot, J. and Jennings, D.H. (1981) *Experimental Mycology*, **5**, 217–28.
38. Hocking, A.D. (1986) *Journal of General Microbiology*, **132**, 269–75.
39. Hocking, A.D. and Norton, R.S. (1983) *Journal of General Microbiology*, **129**, 2915–25.
40. Imhoff, J.F. (1986) *FEMS Microbiology Reviews*, **39**, 57–66.
41. Jakowec, M.W., Smith, L.T. and Dandekar, A.M. (1985) *Applied and Environmental Microbiology*, **50**, 441–6.
42. Jennings, D.H. (1983) *Biological Reviews*, **58**, 423–59.

43. Jennings, D.H. (1984) *Advances in Microbial Physiology*, **25**, 149–93.
44. Jennings, D.H. (1987) *Biological Reviews*, **62**, 215–43.
45. Kauss, H. (1979) *Progress in Phytochemistry*, **5**, 1–27.
46. Kauss, H. (1987) *Annual Review of Plant Physiology*, **38**, 47–72.
47. Kauss, H., Lüttge, U. and Krichbaum, R.M. (1975) *Zeitschrift für Pflanzenphysiologie*, **76**, 109–13.
48. Kouyeas, V. (1964) *Plant and Soil*, **30**, 351–63.
49. Le Rudulier, D. and Boulliard, L. (1983) *Applied and Environmental Microbiology*, **46**, 152–9.
50. Le Rudulier, D., Strom, A.R., Dandekar, A.M., Smith, L.T. and Valentine, R.C. (1984) *Science*, **224**, 1064–8.
51. Le Rudulier, D. and Valentine, R.C. (1982) *Trends in Biochemical Sciences*, **7**, 431–3.
52. Le Rudulier, D., Yang, S.S. and Csonka, L. (1982) *Biochimica et Biophysica Acta*, **719**, 273–83.
53. Lewis, D.H. and Smith, D.C. (1967) *New Phytologist*, **66**, 143–84.
54. Luard, E.J. (1980) *Water Relations of Fungi with Particular Reference to Xerophilic Species*, Ph.D. thesis, Australian National University.
55. Luard, E.J. (1982) *Journal of General Microbiology*, **128**, 2563–74.
56. Luard, E.J. (1982) *Journal of General Microbiology*, **128**, 2575–81.
57. MacRobbie, E.A.C. (1988), in D.A. Baker and J.L. Hall (eds) *Solute Transport in Plant Cells*, Longman Scientific & Technical, London, pp. 453–97.
58. Magan, N. and Lacey, J. (1984) *Transactions of the British Mycological Society*, **82**, 71–81.
59. Mohammed, F.A.A., Reed, R.H. and Stewart, W.D.P. (1983) *FEMS Microbiology Letters*, **16**, 287–90.
60. Nobre, M.F. and D.A. Costa, M.S. (1985) *Canadian Journal of Microbiology*, **31**, 1061–4.
61. Oertli, J.J. (1975) *Zeitschrift für Pflanzenphysiologie*, **74**, 440–50.
62. Papendick, R.I. and Mulla, D.J. (1986), in P.G. Ayres and L. Boddy (eds) *Water, Fungi and Plants*, British Mycological Society Symposium 11, Cambridge University Press, pp. 1–25.
63. Paton, F.M. and Jennings, D.H. (1988) *Transactions of the British Mycological Society*, **91**, 205–15.
64. Pitt, J.I. (1975), in R.B. Duckworth (ed.) *Water Relations of Food*, London, Academic Press, pp. 273–307.
65. Pitt, J.I. and Hocking, A.D. (1977) *Journal of General Microbiology*, **101**, 35–40.
66. Reed, R.H. (1984) *Plant, Cell and Environment*, **7**, 165–70.
67. Reed, R.H., Borowitzka, L.J., Mackay, M.A., Chudek, J.A., Foster, R., Warr, S.R.C., Moore, D.J. and Stewart, W.D.P. (1986) *FEMS Microbiology Letters*, **39**, 51–6.
68. Reed, R.H., Chudek, J.A., Foster, R. and Gadd, G.M. (1987) *Applied and Environmental Microbiology*, **53**, 2119–23.
69. Rosen, B.P. (1986) *Annual Reviews of Microbiology*, **40**, 263–86.
70. Sand, F.E.M.J. (1973), in *Technology of Fruit Juice Concentration — Chemical Composition of Fruit Juices*, Vol. 13, Vienna, International Federation of Fruit Juice Producers, Scientific Technical Commission, pp. 185–216.
71. Slavik, B. (1974) *Method of Studying Plant Water Relations*, New York, Springer Verlag.
72. Smith, L.T. (1985) *Journal of Bacteriology*, **164**, 1088–93.
73. Strøm, A.R., Falkenberg, P. and Landfald, B. (1986) *FEMS Microbiology Reviews*, **39**, 79–86.
74. Sutherland, L., Cairney, J., Elmore, M.J., Booth, I.R. and Higgins, C.F. (1986) *Journal of Bacteriology*, **168**, 805–14.

75. Thornton, J.D., Galpin, M.F.J. and Jennings, D.H. (1976) *Journal of General Microbiology*, **96**, 145–53.
76. Tilbury, R.H. (1980), in G.W. Gould and J.E.L. Corry (eds) *Microbial Growth and Survival in Extremes of Environment*, The Society for Applied Microbiology Technical Series No. 15, London, Academic Press, pp. 103–28.
77. Tilbury, R.H. (1980), in F.A. Skinner, S.M. Passmore and R.R. Davenport (eds) *Biology and Activities of Yeasts*, London, Academic Press, pp. 153–79.
78. Tresner, H.D. and Hayes, J.A. (1971) *Applied Microbiology*, **22**, 210–13.
79. Trinci, A.P.J. and Collinge, A.J. (1974) *Protoplasma*, **80**, 57–67.
80. Troller, J.H. and Christian, J.H.B. (1978) *Water Activity and Food*, London, Academic Press.
81. Truper, H.G. and Galiski, E.A. (1986) *Experientia*, **42**, 1182–6.
82. Van der Walt, J.P. (1970), in J. Lodder (ed.) *The Yeasts: A Taxonomic Study*, Amsterdam, North Holland, pp. 34–113.
83. Von Richter, A.A. (1912) *Mykologisches Zentralblatt*, **1**, 67–76.
84. Warr, S.R.C., Reed, R.H., Chudek, J.A., Foster, R. and Stewart, W.D.P. (1985) *Planta*, **163**, 424–9.
85. Warr, S.R.C., Reed, R.H. and Stewart, W.D.P. (1988) *Plant, Cell and Environment*, **11**, 137–42.
86. Weaste, R.C. (1976) *Handbook of Chemistry and Physics* (57th edn), Cleveland, CRC Press.
87. Wethered, J.M. and Jennings, D.H. (1985) *Transactions of the British Mycological Society*, **85**, 439–46.
88. Wethered, J.M., Metcalf, E.C. and Jennings, D.H. (1985) *New Phytologist*, **101**, 631–49.
89. Wong, P.T.W. and Griffin, D.M. (1976) *Soil Biology and Biochemistry*, **8**, 215–18.
90. Wong, P.T.W. and Griffin, D.M. (1976) *Soil Biology and Biochemistry*, **8**, 219–23.
91. Wyn Jones, R.G. and Pollard, A. (1983), in A. Läuchli and R.L. Bieleski (eds) *Encyclopedia of Plant Physiology, New Series, Vol. 15 Inorganic Plant Nutrition*, New York, Springer Verlag, pp. 528–62.
92. Yagi, T. and Tada, K. (1988) *FEMS Microbiology Letters*, **49**, 317–21.
93. Yancey, P.H., Clark, M.E., Hand, S.C., Bowlus, R.D. and Somero, G.N. (1982) *Science*, **217**, 1214–22.
94. Zimmermann, U. (1978) *Annual Review of Plant Physiology*, **29**, 121–48.

CHAPTER 6

Halotolerant and halophilic microorganisms

D. Gilmour

Introduction

Since it was first recognized in the 1920s and 1930s that microorganisms could grow in habitats containing high concentrations of salt, many microbiologists have attempted to unravel the mechanisms of salt tolerance.[17,18,77,83] Until 1970 the tendency was to concentrate on bacteria, in particular the extremely halophilic bacteria which live in saturated brines. However, more recently the salt tolerance of microalgae has also been examined.[7] During the last fifteen years the mechanisms involved in salt tolerance have been intensively studied and this work has been reviewed at regular intervals.[19,20,78,79,86,111,126] This chapter will consider the salt tolerance of microalgae and bacteria only and the most recent advances in the study of the molecular basis of halotolerance will be examined. Some speculations will be made on the future prospects for work on halophilic/halotolerant microbes and on their possible exploitation for commercial purposes. For information on other halotolerant organisms see ref. 7 for macroalgae, ref. 55 for fungi, refs 61 and 101 for higher plants and ref. 132 for animals.

At this stage a few words need to be said about the terminology that will be used. Any organisms which can grow in the presence or absence of salt is a halotolerant organism. If a requirement for salt is demonstrated, then the organism is a halophile. Two categories of halophiles can be recognized, moderate and extreme halophiles. This terminology will be expanded upon in the section describing the diversity of halophilic/halotolerant microbes.

Table 6.1 Ion contents of the Dead Sea and the Great Salt Lake compared with sea water (g l^{-1})

	Seawater[120]	*Dead Sea*[105]	*Great Salt Lake*[109]
Na^+	10.8	39.2	105.4
K^+	0.4	7.3	6.7
Mg^{2+}	1.3	40.7	11.1
Ca^{2+}	0.4	16.9	0.3
Cl^-	19.6	212.4	181.0
Br^-	0.1	5.1	0.2
SO_4^{2-}	2.7	0.5	27.0
HCO_3^-/CO_3^{2-}	0.1	0.2	0.7
Total salinity	35.2	322.6	332.5
pH	8.2	5.9–6.3	7.7

Habitats

The most extreme natural environments for salt tolerant microbes are the inland lakes (e.g. the Dead Sea and the Great Salt Lake) which are found in sub-tropical or tropical climatic areas.[84,105] These lakes are subject to high rates of evaporation due to the high temperatures and high light intensities which characterize such areas. Highly saline lakes develop when evaporation exceeds the input of freshwater. In addition to natural environments, human activity also creates highly saline habitats for halophilic microorganisms since in many parts of the world salt is produced in solar evaporation ponds or salterns.[78]

The concentration of salts in the Dead Sea and the Great Salt Lake are shown in Table 6.1, where a comparison can be made with seawater concentrations of salts. Both lakes contain high levels of Cl^-, but differ in that the Dead Sea contains a higher level of Mg^{2+} and a lower level of Na^+ than the Great Salt Lake. It is often found that microbes isolated from the Dead Sea require high levels of Mg^{2+} for good growth.[85] The pH of the Dead Sea is slightly acidic and that of the Great Salt Lake slightly alkaline, but highly saline environments exist which are also highly alkaline, e.g. soda lakes found in the Rift Valley of Kenya (ref. 54 and Chapter 3 in this book). The different composition of these saline lakes appears to depend on the relative abundance of Ca^{2+} and Mg^{2+} in the surrounding rocks. These cations influence the final composition by removing certain ions as insoluble salts. Grant and Ross[54] suggest that a high Ca^{2+} area will produce a slightly alkaline lake dominated by Na^+ and Cl^- (e.g. Great Salt Lake) whereas high Ca^{2+} and Mg^{2+} areas produce slightly acidic Mg^{2+}, Na^+ and Cl^- dominated brines (e.g. Dead Sea).

In addition to these extremely saline environments, extreme habitats are found in rather less exotic places, e.g. in estuaries and particularly in shoreline rockpools. When the tide is out, rockpools are exposed to evaporation and on warm days the salinity increases several-fold, and occasionally the salinity can approach the level found in salt lakes. The incoming tide then causes a rapid decrease in the salinity and thus rockpool organisms must be able to withstand sudden large decreases in salinity as well as tolerating high salt levels.[96]

The organisms

Table 6.2 lists the major types of microalgae and bacteria found in saline habitats. They can be classified as follows:

halotolerant Optimum growth between 0 and 0.3 M NaCl and a growth range 0–1 M NaCl.
moderately halophilic Optimum growth between 0.2 and 2.0 M NaCl and a growth range 0.1–4.5 M NaCl.
extremely halophilic Optimum growth between 3.0 and 5.0 M NaCl and a growth range 1.5–5.5 M NaCl.

A moderate halophile can be distinguished from a halotolerant organism by having a requirement of at least 0.1 M NaCl for growth, and they can be separated from the extreme halophiles by never requiring as much as 1.5 M NaCl for growth. Therefore, in this system the minimum level of NaCl which will support growth is one of the major parameters. Some halotolerant microbes also require Na^+, but only in the micromolar range. It should also be noted that an organism's response to salt will depend on other environmental factors (temperature, pH, etc.), thus care must be exercised in assigning organisms to these categories.

The halotolerant group mainly consists of the enteric bacteria plus a wide variety of microalgae (Table 6.2). Gram-negative bacteria belonging to the genus *Halomonas* are also placed in this group and *H. elongata* is the most halotolerant species isolated thus far, growing well at salinities from 0.05 M up to saturated NaCl (5.5 M). The only organism which comes close to equalling this salinity range is the unicellular green alga *Dunaliella viridis*, which has a salinity range 0.3–5.0 M NaCl (Table 6.2). *Dunaliella* is unique because species from this genus are found in all three categories of salt tolerance shown in Table 6.2. However, no one species of *Dunaliella* can grow over the entire range of salinities shown by *H. elongata*. It is generally assumed that halotolerant microbes grow best at the lower end of the salinity scale, but surprisingly few data exist to demonstrate that this is true.

The moderate halophiles contain representatives from the eubacteria, cyanobacteria and microalgae. Three genera of obligately anaerobic fermentative bacteria isolated from the Dead Sea and the Great Salt Lake are also included. The optimum salinity for growth is normally considerably less than their upper salinity limit for growth. Thus, moderate halophiles can tolerate high salinities but do not grow optimally in them. The extreme halophiles consist of only a few highly specialized groups. Predominant are the halobacteria which belong to the archaebacteria,[131] *Halobacterium* and *Halococcus* are found in slightly acidic or slightly alkaline conditions typical of the Dead Sea and the Great Salt Lake, whereas *Natronobacter* and *Natronococcus* are found in the highly alkaline conditions

Table 6.2 Halophilic/halotolerant microorganisms

	Salinity range (M) (optimum)	*Comments*	*Reference*
HALOTOLERANT			
Bacteria			
Enteric bacteria (*E. coli*, *Salmonella*, *Klebsiella*)	0–1.0 (0.2–0.3)	best understood group at the molecular level	20
Halomonas halodurans	0–3.0		126
Halomonas elongata	0.05–5.5	has widest range of salt tolerance known	126
Planococcus	0–1.5	also psychrotolerant	98
Algae			
Chlamydomonas pulsatilla	0–1.5		62
Chlorella autotrophica	0–2.0		62
Chlorella emersonii	0.01–0.33		7
Cyclotella	0–0.5		62
Dunaliella tertiolecta	0.025–3.5		87
Monallantous	0.04–0.6		7
Nannochloris	0–1.5		62
Platymonas	0.05–1.5		62
Stichococcus	0–0.7		62
MODERATELY HALOPHILIC			
Bacteria			
Anaerobic bacteria (*Halobacteroides*, *Sporohalobacter*, *Haloanaerobium*)	1.0–3.0 (1.5–2.0)	degrade a wide variety of organic substrates	106
Ba_1 (Gram-negative)	0.2–4.0 (0.8)		71
Micrococcus	0.2–4.0 (1.0–2.0)		126
Synechococcus	0.5–4.0	includes strains previously designated *Aphanothece* or *Aphanocapsa*	111
Vibrio alginolyticus	0.2–1.5		65
Vibrio costicola	0.5–3.3 (0.8–1.5)		1
Algae			
Astereomonas	0.5–4.5		7
Botryococcus	0.4–2.8		7
Chlamydomonas sp.	0.34–1.71		7

Table 6.2 – cont.

	Salinity range (M) (optimum)	*Comments*	*Reference*
Chlorococcum	0.1–2.0 (0.25–0.375)	high lipid content	Blackwell and Gilmour, unpubl.
Dunaliella parva 19/9	0.1–2.0	plus other marine *Dunaliella* species	49
Dunaliella viridis	0.3–5.0	widest range known for a eukaryote	13
Navicula	0.25–2.75		7
Pavlova	0.15–1.0	originally called *Monochrysis*	111
Phaeodactylum	0.14–0.77		7
Porphyridium	0.25–1.0	range is for more than one species	125
EXTREMELY HALOPHILIC			
Bacteria			
Ectothiorhodospira	1.5–5.0 (3.4)	phototroph	20, 40
Halobacteria (*Halobacterium*, *Halococcus*, *Natronobacter*, *Natronococcus*)	2.0–5.5 (3.5–5.0)		20
Algae			
Dunaliella 'D13'	2.0–5.5	plus several more species	49, 62

of the Kenyan soda lakes.[54] The only eubacterium in this group is the phototroph *Ectothiorhodospira* which also flourishes in soda lakes,[54] and the only eukaryotes are species of *Dunaliella* (Table 6.2).

In the next section the physiology of two of the organisms listed in Table 6.2, *Halobacterium* and *Dunaliella*, will be examined in detail. Both of these organisms have been intensively studied and between them these genera encompass many of the properties of halophiles since *Halobacterium* is an extreme halophile whereas most species of *Dunaliella* can be classified as halotolerant or as moderate halophiles. In addition, the moderately halophilic bacteria will be examined, since some aspects of halotolerance have only been studied in this group. Although, the physiology of salt tolerance is generally less well understood in moderately halophilic bacteria.

Halotolerance

In halophiles, it has been shown that the internal concentration of NaCl does not equal the external concentration at high salinities, i.e. as the external NaCl concentration is increased, mechanisms come into play which prevent a concomitant increase in intracellular NaCl.[19] High internal concentrations of Na^+ are toxic to cells and cell membranes therefore have lower permeabilities to Na^+ than K^+, which is much less toxic. Na^+ is often actively extruded from cells.[111] Thus most living cells contain high levels of K^+ and low levels of Na^+. However, in halophiles, in addition to K^+ accumulation, it is necessary to accumulate other osmotically active compounds to allow volume (in wall-less cells) or turgor (in walled cells) to be regulated. Since these compounds were found to be less injurious than NaCl to the structure and function of macromolecules and enzymes, and in fact many could actively protect against NaCl inhibition of cellular processes, the term *compatible solutes* was coined for them by Brown.[19]

The vast majority of halophiles use small molecular weight organic compounds as compatible solutes and this allows the exclusion of most (if not all) of the external NaCl. This is the mechanism used by *Dunaliella*. In contrast, in the halobacteria, e.g. *Halobacterium*, KCl is accumulated as the compatible solute and NaCl is also accumulated, although to a lesser extent than KCl. Therefore, these two genera illustrate well the two fundamental mechanisms of salt tolerance and more detailed consideration of these mechanisms follows.

DUNALIELLA

In 1964, Craigie and McLachlan[27] found that glycerol was the major compound synthesized by *Dunaliella* as measured by the incorporation of $^{14}CO_2$ and external salinity had an effect on the amount of glycerol synthesized. However, it was not until the early seventies that Wegmann[128] and Ben-Amotz and Avron[4] independently demonstrated that the intracellular concentration of glycerol in *Dunaliella* increased with rises in external salinity. Borowitzka and Brown[12] demonstrated the compatible solute nature of glycerol in *Dunaliella* by showing that glycerol concentrations up to 4 M do not inhibit the activity of glucose-6-phosphate dehydrogenase isolated from *D. tertiolacta* or *D. viridis*. Glycerol may indeed show some degree of activation of the enzyme at concentrations around 3 M. In contrast, 0.5 M NaCl plus 0.5 M KCl caused a 60% inhibition of *Dunaliella* glucose-6-phosphate dehydrogenase and double this amount of salt caused complete inhibition.[12] Therefore, the mechanism of adaptation to increased salinity in *Dunaliella* seems straightforward.

1. Immediately the external NaCl concentration is increased, water flows from the cell causing a decrease in volume (due to the lack of a cell wall).
2. The decrease in volume triggers the synthesis of glycerol which then takes place over the next 6–8 hours.
3. The cell volume recovers to approximately the level before the salinity increase and growth recommences with a higher concentration of glycerol inside the cells.[6]

Exactly the opposite reactions occur when the cells are subjected to salinity decreases.

INHIBITION OF PHOTOSYNTHESIS

The increase in external salinity, which induces a decrease in cell volume, also inhibits the rate of photosynthesis. By measuring the absorption of light at 519 nm it is possible to estimate the magnitude of the charge separation across the thylakoid membranes which is used to drive the photosynthetic production of ATP. Immediately after a salinity increase, it was shown that ions (presumably Na^+ or Cl^-) enter the chloroplast and abolish the charge separation across the thylakoid membranes and it is this that inhibits the rate of photosynthesis.[41] Large increases in the external salinity can completely inhibit photosynthesis for 12 hours or more, but the cells can still synthesize glycerol during this period.[45] Thus, it was clear that an alternative source of glycerol had to be available other than that from CO_2-fixation. Gimmler and Moller[45] showed this source to be starch (the storage product of the alga) and glycerol can be produced by breakdown of starch while photosynthesis is inhibited.[41] As well as allowing the cell volume to return to normal, the production of glycerol also enables the cells to remove the excess intracellular Na^+ and Cl^- ions, thereby relieving the inhibition of photosynthesis. Earlier the toxicity of Na^+ ions to cellular function was stressed, but it is often overlooked that Cl^- ions are also inhibitory to many organisms. In addition to a possible role in the salt-induced inhibition of photosynthesis, Gimmler and co-workers showed clearly that a range of enzymes isolated from *D. parva* were inactivated by Cl^- ions and not by Na^+ ions.[46]

RESPONSE TO DECREASED SALINITY

When the external salinity is decreased, glycerol is converted back to starch; if the decrease is severe, then glycerol may be excreted into the medium.[42] Recent work by Zidan *et al.*[133] showed that both processes of glycerol removal were inhibited by the uncoupler carbonylcyanide *m*-chlorophenyl hydrazone (CCCP), indicating a dependence on ATP availability for both metabolic conversion of glycerol to starch and for glycerol excretion. The observation that glycerol extrusion is an active process allows the reconciliation of two apparently contradictory factors, i.e. that glycerol leaks from *Dunaliella* cells[39] and that the alga has an unusual plasma membrane with a very low permeability coefficient for glycerol possibly due to the presence of sterol peroxide lipids as major constituents of the membrane.[22,119] Therefore, it appears that *Dunaliella* cells do have a plasma membrane which is largely impermeable to glycerol, and it is necessary to expend energy to excrete glycerol into the medium when a severe salinity decrease is imposed on the cells.

Despite the ability of the cells rapidly to alter the internal concentration of glycerol, the immediate response to a decrease in salinity is a rapid increase in cell volume. When the salinity decrease is severe, the increase in cell volume can require almost instantaneous increases in cell surface area of up to 400%.[30] Maeda and Thompson[92] examined the mechanism which allowed the cells to cope with the need for such rapid membrane synthesis and they found that the plasma membrane expanded by fusing with small vesicles (about 0.25 μm in diameter) which are prevalent in the cytoplasm. In addition to plasma membrane expansion, it was found that membranes surrounding organelles also expanded in surface area and, in the case of the chloroplast envelope, the new membrane came from the endoplasmic reticulum.[92]

GLYCEROL CYCLE

The enzymes involved with glycerol synthesis and degradation have been identified, purified and characterized, and the localization of some within the cell has been determined (e.g. refs 44, 60, 95). The pathway is known as the glycerol cycle, with glycerol being produced from dihydroxyacetone phosphate via glycerol phosphate and being removed via dihydroxyacetone back to dihydroxyacetone phosphate (Fig. 6.1). The localization data suggest that glycerol-3-phosphate dehydrogenase and at least some of the glycerol-3-phosphatase are in the chloroplast, which means that glycerol synthesis from dihydroxyacetone phosphate can take place solely in the chloroplast. The localization of dihydroxyacetone kinase is not known, but the oxidoreductase enzyme is in the cytoplasm suggesting that glycerol is degraded outside the chloroplast (ref. 44 and Fig. 6.1). Recently, the turnover rate of glycerol in *Dunaliella* cells was determined and found to be rather slow (up to 6 hours for half time of turnover). This has led to the suggestion that glycerol metabolism resembles that of storage products rather than a conventional metabolic cycle.[52]

SENSING SALINITY CHANGE

One question which must be answered before a complete explanation of halotolerance in *Dunaliella* can be presented is what is the 'trigger' mechanism which switches on glycerol synthesis after an external salinity increase? Clearly, the reduction in volume concentrates all the metabolites within the cell. Gimmler[45] suggested that inorganic phosphate concentration may be involved in the signal mechanism; this has been shown to be the case for controlling the balance between starch and glycerol within the cell.[53] Recently Belmans and Van Laere[3] have suggested that inorganic phosphate may exert its effect at the level of glycerol-3-phosphatase (Fig. 6.1), since it is a competitive inhibitor of this enzyme. They have calculated that during steady-state conditions there is insufficient levels of glycerol-3-phosphate (0.6 mM) to allow much glycerol synthesis due to the presence of 1 mM inorganic phosphate. However, the same authors have shown that there is increased activity of glycerol-3-phosphate dehydrogenase immediately after a salinity increase. This leads to a higher level of glycerol-3-phosphate (5 mM) which is sufficient to overcome the competitive inhibition by inorganic phosphate and glycerol is synthesized.[2] The plasma membrane may be involved, pressure changes may induce a 'message' to be sent from the plasma membrane to increase the activity of glycerol-3-phosphate dehydrogenase and thus initiate glycerol production.[30] This is still speculative, and it is obvious that more evidence is needed before the signal mechanism can be identified.

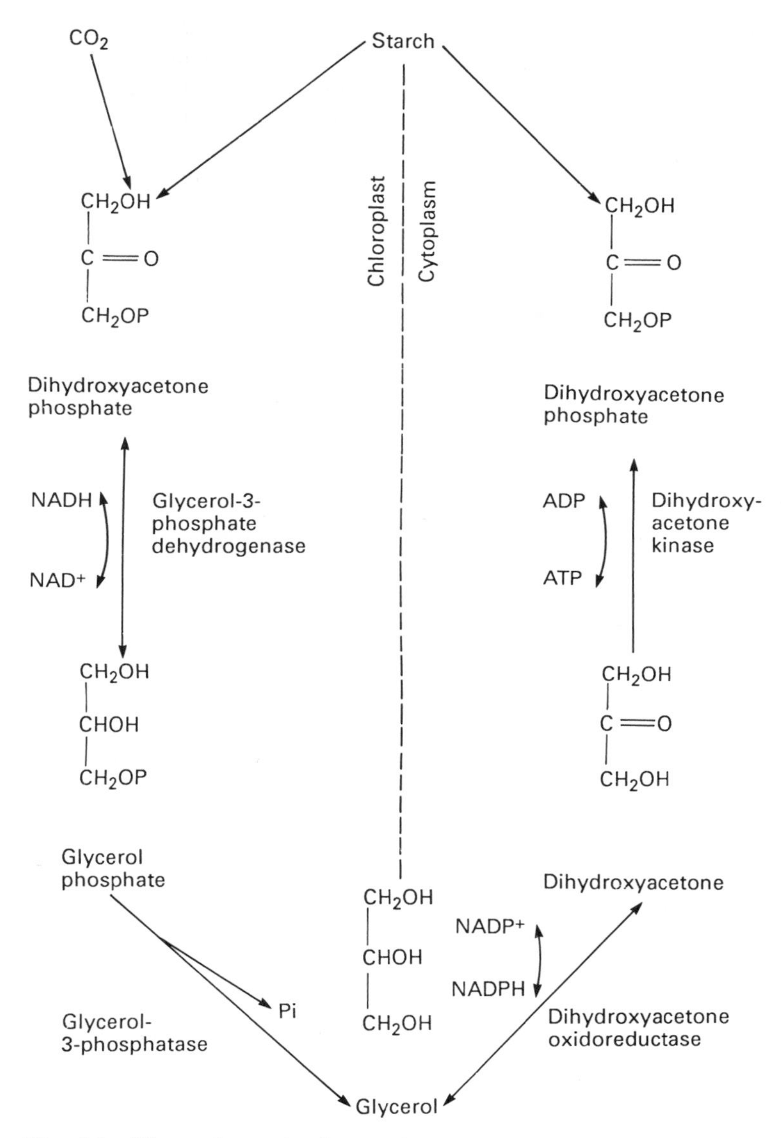

Fig. 6.1 The pathways leading to the production and removal of intracellular glycerol in *Dunaliella* (adapted from ref. 6).

ROLE OF INORGANIC IONS

The second problem involves the degree to which glycerol accounts for the balancing of internal osmotic pressure with elevated external osmotic pressure. At first Ben-Amotz and Avron believed that glycerol accounted for virtually all of the internal osmotic pressure and that the internal ion concentrations were very low.[6]

However, work from the same laboratory has now modified the figure to about 83% of the osmotic balance considered to be due to glycerol, with the rest due to other soluble organic metabolites.[31] Recently evidence has been gathered by Avron and co-workers, using new methods of measuring cell volume and by looking at ion fluxes in isolated plasma membrane vesicles, to support their case for a low ionic environment within *Dunaliella* cells brought about by carrier mediated transport at the plasma membrane.[69,108]

Nevertheless, other groups believe that, although the level of intracellular glycerol is highly significant, complete balance is only achieved due to the presence of significant amounts of Na^+ and Cl^- ions within the cells.[43,50] As already mentioned, it is certainly true that, immediately after a salinity increase, the intracellular concentrations of Na^+ and Cl^- increase.[48,56] The point in question is whether significant amounts of intracellular Na^+ and Cl^- ions are found in cells adapted to high salinities. A possible answer to this question lies in the recent observations of vacuoles in *Dunaliella* cells[56] confirming earlier observations[64] which suggest that significant amounts of Na^+ and Cl^- could be concentrated in these vacuoles in cells adapted to high salinities.

To conclude, the mechanism of salt tolerance in *Dunaliella* is well established and glycerol synthesis and degradation clearly play a major role. However, the 'trigger' mechanism for sensing external osmotic pressure and switching glycerol synthesis on and off is not fully understood, and the role (if any) of internal Na^+ and Cl^- ions in balancing the external osmotic pressure requires further elucidation.

OTHER HALOTOLERANT AND HALOPHILIC ALGAE

With the exception of a few species of *Dunaliella*, algae which grow in elevated salt concentrations can either be halotolerant or moderately halophilic (Table 6.2). Only a few algae can grow well in high salinities, e.g. *Astereomonas*, a prasinophyte alga, has almost as wide a salinity range as *Dunaliella* and it also uses glycerol as a compatible solute. Another prasinophyte, *Botryococcus*, has been reported to grow in 3.0 M NaCl.[7] *Chlamydomonas*, a green alga closely related to *Dunaliella*, has some strains which can grow in salinities up to 1.7 M NaCl. One or two species of diatoms appear to be moderate halophiles, e.g. *Navicula*. However, with the exception of *Dunaliella*, none of the moderately halophilic algae have been studied in any great detail.[7]

Among the halotolerant algae (i.e. those algae which do not require substantial amounts of NaCl), species of *Chlamydomonas*, *Chlorella*, *Nannochloris* and *Platymonas* can grow in salinities in the region of 1.5–2.0 M NaCl. The other algae found in this group have a more restricted salinity range, not growing well in salinities much above seawater levels (approximately 0.5 M NaCl). All of the halotolerant algae listed in Table 6.2 have been studied to some extent and the most common compatible solutes used are glycerol, proline and mannitol.[7,62] Again, none of these species has been as well characterized as *Dunaliella*, although the prasinophyte alga *Platymonas* has been extensively investigated by Kirst (e.g. refs 73, 74). In *Platymonas*, mannitol is accumulated in response to salinity increases; the synthesis of mannitol is not light dependent and it takes place in the presence of electron transport inhibitors or uncouplers (which abolish the electrochemical H^+ gradient), albeit after a lag of 10–15 min and at a slower rate. *De novo* synthesis of enzymes is not necessary for mannitol accumulation to take place. Immediately after an external salinity change, major intracellular ions (particularly K^+) are involved initially in balancing the osmotic potential of the cells to allow time for mannitol synthesis or degradation to occur.[73,74] However, when the sum of the osmotic potential of inorganic ions and mannitol was calculated, it was found to account for only 60% of that required for osmotic balance with the external medium.[72] Subsequently three other compatible solutes were identified in *Platymonas*, the tertiary sulphonium compound dimethylsulphoniopropionate and the quaternary ammonium compounds betaine and homarine, which appear to make up the missing osmotic potential.[34]

HALOBACTERIUM

The mechanism of salt tolerance in *Halobacterium* is very different from the one outlined above for *Dunaliella*. In fact, tolerance is misleading in this context, since *Halobacterium* requires at least 3.0 M NaCl for optimum growth. The key to understanding how *Halobacterium* grows in high salt is to realize that they are found in extreme but stable environments, unlike some species of *Dunaliella* which live in areas of fluctuating salinity. In consequence, *Halobacterium* has become irreversibly adapted to high salt environments.

Halobacterium belongs to the archaebacteria which, unlike eubacteria, have no muramic acid in the cell wall and have ether-linked rather than ester-linked lipids. Two other properties, however, are of direct interest in terms of their salt relations:[20,78] their cytoplasmic proteins and ribosomes are highly acidic and their membrane lipids are predominantly acidic. It seemed clear that such proteins and lipids would require high levels of cations to screen the negative charges and stop repulsion forces pulling the molecules apart. When the internal milieu of *Halobacterium* was investigated,[25] a high salt concentration was found but, instead of NaCl equimolar with that outside, a substantial amount of internal salt was KCl (4.0 M) with NaCl (0.7 M) a significant but more minor component.[51]

ENZYME ACTIVITY

The activities of enzymes extracted from *Halobacterium* were measured to see if they varied with salt content and composition.[78,80] Two major types of enzyme response to salt were found (Fig. 6.2):

1. those such as aspartate transcarbamylase which require 2 M salt for activity and have an optimum around 4 M.
2. those such as isocitrate dehydrogenase which function best at salinities around 1 M, and are inhibited by higher concentrations of salt.

A third type represented by fatty acid synthetase show little activity at salt concentrations above 1 M and must be largely inactive *in vivo*.[82] Kushner[78] states

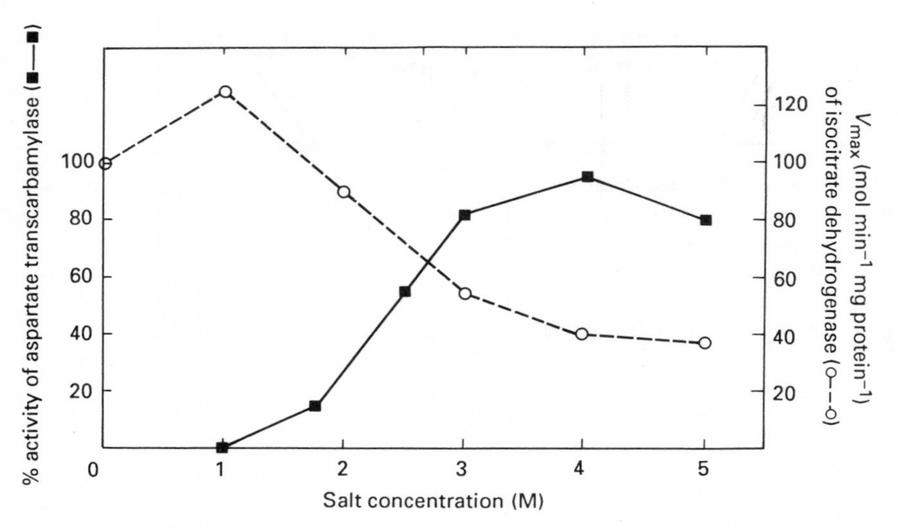

Fig. 6.2 Salt dependence of two enzymes isolated from *Halobacterium* (adapted from ref. 82).

that about half the enzymes studied have approximately the same activity in NaCl and KCl; however, others shown more activity in KCl than NaCl. This appears to be due to K^+ binding much less firmly to enzymes than Na^+.[19,20] It should be noted that high salt levels stabilize halophilic enzymes, but the activity under these conditions may be low, e.g. isocitrate dehydrogenase has little activity in 4 M salt (Fig. 6.2) but its stability is best at this salt concentration.

In addition to NaCl and KCl, other salts such as NH_4Cl and LiCl can also activate halobacterial enzymes.[78] The highest activity for the malic enzyme extracted from *Halobacterium* was found in NH_4Cl rather than KCl.[124] However, the allosteric regulation of this enzyme was more effective in KCl, making it the more physiologically appropriate solute.

ADAPTATION OF PROTEINS AND LIPIDS

In 1974, Lanyi suggested that there is more to the interaction of Na^+ and K^+ with the acidic proteins and lipids than simple screening of charges, since much smaller amounts of these cations would be sufficient.[82] It was also observed that there are fewer non-polar amino acids in *Halobacterium* proteins than are found in the proteins of other organisms, which has the consequence of weakening the hydrophobic interactions within them. Thus, *Halobacterium* requires K^+ at intracellular concentrations of at least 3 M to take its proteins out of solution and allow the correct hydrophobic bonds to be made in order to achieve the active configuration of the protein.

Current views on this subject concur with the above and emphasize that, as well as an overall excess of acidic over basic residues, there are acidic clusters in sequenced proteins from *Halobacterium*. These clusters may protrude out into the

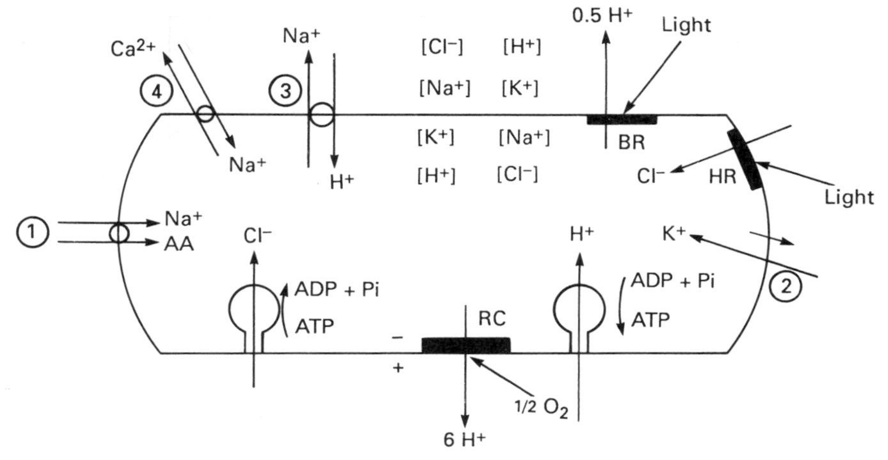

Fig. 6.3 Active transport systems of *Halobacterium*, see text for references. RC = respiratory chain, HR = halorhodopsin; BR = bacteriorhodopsin. 1, active uptake of amino acids coupled to Na^+ entry (symport); 2, K^+ entry in response to the electrochemical potential; 3, active Na^+ extrusion is coupled to H^+ entry (antiport); 4, Ca^{2+} extrusion is coupled to Na^+ entry (antiport).

solvent.[38,129] In addition, halophilic proteins have the ability to retain large amounts of water and salt. Harel *et al.*[58] have crystallized malate dehydrogenase from *H. marismortui* and found there was 0.87 g water g protein^{-1} and 0.35 g NaCl g protein^{-1}. A typical non-halophilic protein, bovine serum albumin, has 0.23 g water g protein^{-1} and 0.012 g NaCl g protein^{-1}. It has been proposed that stability in high salinity arises from competition for water between the negatively charged protein residues and external salt.[38,129]

ION PUMPS AND MEMBRANE TRANSPORT

Halobacterium is normally a strict aerobe with chemoheterotrophic nutrition, although some species can synthesize ATP from light driven proton (H^+) pumping by the 'purple membrane' pigment bacteriorhodopsin. The protons are pumped through the cell membrane and thereby generate an electrochemical potential.[121] ATP synthesis due to bacteriorhodopsin activity could take place under anaerobic conditions, but this is unlikely to be a significant growth strategy in *Halobacterium*, since the synthesis of the retinol base of bacteriorhodopsin requires oxygen. A second pigment, halorhodopsin, has been shown to act as a light driven inwardly directed Cl^- ion pump.[37] The cytoplasmic concentration of Cl^- in *Halobacterium* is approximately equal to that in the external medium; however, when the membrane potential is taken into account (90–150 mV, inside negative), there is a role for active transport of Cl^-. In addition, a second Cl^- uptake system was shown to be ATP dependent (Fig. 6.3). Since it was known that intracellular Na^+ levels were kept much lower than the external concentration, a primary pump for Na^+ extrusion was sought. Indeed, halorhodopsin was initially thought to be this pump, before being assigned the role described above.[121] No Na^+ pump has been found and it appears that Na^+ extrusion takes place as a consequence of H^+ and Cl^- active transport by bacteriorhodopsin and halorhodopsin, and respiration mediated ion movements (Fig. 6.3). Therefore, the H^+ gradient imposed by bacteriorhodopsin and respiratory activity acts in an exchange (antiporter) mechanism with Na^+, which removes Na^+ from the cell at the expense of abolishing part of the H^+ gradient.[76,103] The Na^+/H^+ antiporter has been shown to be sensitive to N,N′-dicyclohexylcarbodiimide (DCCD) and the addition of monensin (electroneutral Na^+/H^+ exchanger) restored the extrusion of Na^+.[102] The concentration gradient of Na^+ which is maintained between the cell and the external medium drives the transport of amino acids into the cell and the extrusion of Ca^{2+} from the cell. The various ion pumps and the gradients set up are shown in Fig. 6.3.

DIVERSITY WITHIN THE GENUS

No attempt has been made here to differentiate between species of *Halobacterium* because of the lack of biochemical diversity which is presumably due to the constraints of life in high salt environments. This means that classification of halobacteria on a phenotypic basis is difficult. As an alternative, Ross and Grant[115] used nucleic acid hybridization techniques to examine a wide variety of halobacterial strains. Within the genus *Halobacterium*, the main conclusion was that *H. halobium* and *H. cutirubrum* should be considered identical to *H. salinarium*. Despite the lack of phenotypic variation, evidence is growing that at least two species within the genus have very different characteristics. *H. marismortui*, which was isolated from the Dead Sea, appears to have a much more permeable plasma membrane than *H. salinarium*. This was first observed by Ginzburg[47] almost twenty years ago when she found that *H. marismortui* was freely permeable to compounds as large as inulin (molecular weight about 5000). These results were received with some scepticism, due to the problems of reconciling them with membrane properties needed for growth. However, more recently the Ginzburg group have used dielectric measurements to confirm their earlier findings[100] and it seems clear that, even within the extremely halophilic bacteria, *H. marismortui* is a very unusual organism worthy of further study.

MODERATELY HALOPHILIC BACTERIA

This heterogeneous group of microbes are much less well understood than the extreme halophiles. As a group they can adapt to a wide range of salinities, 0.2–4.0 M. Organic-compatible solutes are found in these bacteria; betaine is the most widespread and is obviously important for halotolerance.[65] Nevertheless complete osmotic balances have rarely been reported for moderately halophilic bacteria and the internal concentrations of inorganic ions are uncertain.

ENZYME ACTIVITY

Exoenzymes (nucleases and amylases) from moderately halophilic bacteria are not surprisingly halophilic (i.e. depend on high concentrations of salt for maximal activity), as are some membrane-bound enzymes such as 5′-nucleotidase from *Vibrio costicola* which is in contact with the external medium. Some intracellular enzymes from *V. costicola*, such as pyruvate kinase, function well in the absence of salt or in the presence of high salt concentrations. However, other intracellular enzymes from the same organism (e.g. threonine deaminase) are inhibited in the presence of 3 M NaCl or KCl;[80] these are salt concentrations which could be present if organic-compatible solutes do not play a significant role in *V. costicola* (Table 6.2). Kushner[81] has suggested that salt-induced inhibition of enzyme activity may be due to Cl^- toxicity, since in the cell-free enzyme assays Cl^- has invariably been used as the balancing anion for Na^+ and K^+. There is some evidence that the intracellular level of Cl^- is kept low and this may be essential for enzyme activity, rather than a low salt environment *per se*. The same appears to be true for protein synthesis.[80] Unless another anion besides Cl^- is present in large amounts (which seems unlikely), this would imply that the vast majority of cations are associated with the surface layers of the cells.[81] If this hypothesis is confirmed it would be a fundamental difference between extreme and moderate halophiles, since extremely halophilic bacteria do not maintain a low level of Cl^- within their cells (Fig. 6.3).

PHOSPHOLIPID SYNTHESIS

The response of moderately halophilic bacteria to increased salinity has been studied by Kogut, Russell and co-workers using *Vibrio costicola*.[116] Phospholipid synthesis was shown to alter when the cells were transferred into higher salinity; the proportion of negatively charged phospholipids to neutral (zwitterionic) phospholipids increased. Immediately after the salinity increase all phospholipid synthesis was inhibited, but that of phosphatidylethanolamine (neutral lipid) more so. During the lag phase before growth recommenced, the synthesis of phosphatidylglycerol and diphosphatidylglycerol (negatively charged phospholipids) increased more rapidly than phosphatidylethanolamine. This led to the altered ratio and the higher level of negatively charged phospholipids was maintained during steady-state growth in high salinity.[57,116] The altered ratio of phospholipids was also induced by increasing the external sucrose concentration, thus it appears that changes in the external osmotic potential influence the lipid composition of *V. costicola* and this may be the basis of a sensing mechanism which allows the cells to adapt to altered external salinity.[1]

ACTIVE TRANSPORT AND ENERGY GENERATION

Another aspect of the physiology of moderately halophilic bacteria which has been studied in detail is the active transport mechanisms which, it is proposed, keep the intracellular Na^+ level lower than that in the external medium. *Vibrio alginolyticus* possesses a primary Na^+ extrusion system (Na^+ pump) which is coupled to respiration.[122] Tokuda and co-workers have shown that the Na^+ pump is coupled to a NADH oxidase which requires Na^+ for maximum activity. Uncoupler studies using carbonycyanide *m*-chlorophenylhydrazone (CCCP), which abolishes the electrochemical H^+ gradient set up by respiration, have shown that a membrane potential can still be generated by *V. alginolyticus* in the presence of CCCP at pH 8.5. The authors concluded that the NADH oxidase translocates Na^+ as a direct result of respiratory activity and thus generates a Na^+ electrochemical potential.[123] Similar results showing a respiration linked, uncoupler insensitive Na^+ pump have also been reported for another moderately halophilic bacterium designated Ba_1.[70] Thus, in contrast to the extremely halophilic bacteria, direct Na^+ pumps may be widespread in moderate halophiles.

Therefore, there is good evidence that the internal salt concentration of moderate halophiles is lower than the external concentration. This is in agreement with the data showing significant amounts of compatible solutes (particularly betaine) in this group of bacteria.[65] Nevertheless, the relative importance of organic compatible solutes in moderately halophilic bacteria remains to be fully quantified.

GENERAL STRATEGY FOR HALOPHILES

The mechanism of salt tolerance already described for *Dunaliella* can be generalized to postulate a strategy for all halophiles except the halobacteria.

1. Immediately the external salinity is increased, walled cells take up K^+ from the medium to restore cell turgor pressure; in wall-less cells the cell volume decreases.
2. The accumulation of K^+ (or some other intracellular constituent in wall-less cells) acts as a signal for synthesis or uptake of the compatible solute(s). The intracellular K^+ content decreases again and in wall-less cells the cell volume returns to normal.

A reversal of this mechanism occurs when the external salinity is decreased.

There is good evidence for this scheme in halotolerant strains of *E. coli* and *Salmonella*.[63] Is a similar strategy found throughout halotolerant microbes and moderate halophiles? At the moment this is uncertain, since our understanding of the role of inorganic ions in many species is poor; this is particularly true for the moderately halophilic bacteria.

Halotolerant higher plants also have much in common with this strategy since they regulate ion levels in the cytoplasm and accumulate similar organic compatible solutes to halotolerant microorganisms to make up the cytoplasmic osmotic balance. However, in higher plants, more than 80% of the cell volume is taken up by the vacuole and it is clear that solutes present in the vacuoles are inorganic ions.[101] The giant-celled algae (e.g. *Valonia*) have large vacuoles and they could be used as a model system for investigating higher plant salt tolerance; however, they are marine algae and are not particularly salt tolerant.[61]

PROPERTIES OF COMPATIBLE SOLUTES

The nature and properties of compatible solutes has been alluded to already in this chapter and they have been extensively reviewed.[11,19,20,91,132] Not all compatible solutes are equally effective; in general, betaine and glycerol are better than proline which in turn is better than glutamate. Compatible solutes must not perturb macromolecules at physiological pH (around 7); they need to be negatively charged or neutral (often zwitterions), but not positively charged. There appears to be a limit to which K^+ can be accumulated in response to salt stress, which is presumably due to the need to balance its positive charge with organic anions such as malate or glutamate. Above this level a more compatible solute must be synthesized to take the place of K^+. The only known exception to this rule is the halobacteria, which are highly specialized in a way that other halophiles are not. K^+ is favoured over other cations such as Na^+ or Li^+ because it tends to alter water structure in a way which stabilizes active enzyme conformation rather than inducing solubility, i.e. salting out rather than salting in. Another requirement, if a compound is to be a compatible solute, is that it must be retained by the cell membrane often against very large concentration gradients.

The series of compounds leading from glycine to betaine through sarcosine (methylglycine) and dimethylglycine have been investigated in terms of their compatible solute activity. The level of compatible solute activity increases with the increasing methylation of the compounds the highest activity being found in the trimethylated betaine.[91] Sutton and Lilley (personal communication) have recently shown that this is due to the increased number of methyl groups screening the positive charge on the nitrogen. This makes the molecule less reactive and thus less perturbing. The same authors have found that betaine and proline, two common compatible solutes, have ideal osmotic coefficients greater than one. This means that less of the compound need be produced to achieve osmotic balance with the external medium, making these compatible solutes more energetically attractive than was first thought. Borowitzka[11] pointed out that some solutes such as ethylene glycol and threitol satisfy all the requirements to be compatible solutes, but have not been demonstrated to act this way in living cells. Two possible reasons for this are (i) that there are some biochemical restrictions to their synthesis by cells or (ii) that there are further properties necessary to be a compatible solute which are not clear to us at present.

COST OF SALT TOLERANCE MECHANISMS

The two contrasting mechanisms of salt tolerance shown by *Dunaliella* and *Halobacterium* impose different 'costs' on the cells. In the case of *Dunaliella*, there is an energetic cost in maintaining high concentrations of the compatible solute glycerol within the cells, especially in high salinities where the production of glycerol accounts for much of the fixed CO_2. These energetic costs are not so severe for *Halobacterium*, although energy must be expended in maintaining the large opposing gradients of K^+ and Na^+ across the cell membrane (Fig. 6.3). On the other hand, the high internal salt concentration of *Halobacterium* has led to irreversible changes in macromolecular structure which lead to complete loss of activity at 'normal' salinities. It can be argued that this is a cost to *Halobacterium* in terms of flexibility, i.e. they cannot colonize habitats other than hypersaline brines.

The last point to make in this section is the cost in terms of molecular biology. The number of mutations needed to produce a cell capable of using the *Halobacterium* strategy is high, since the majority of enzymes and macromolecules differ from their low-salt-adapted equivalents. However, this is not true for the *Dunaliella* type strategy; in this case, a few mutations could transform a non-halophile into a halotolerant organism.

Commercial aspects

Two main possibilities exist for the commercial exploitation of halophiles, which reflect the two contrasting strategies of salt adaptation described above. The first possibility is to use halophiles which produce compatible solutes in large amounts as a source of useful chemicals, which could form the basis of industrial processes. The second possibility concerns those halophiles which do not synthesize organic compounds as compatible solutes but instead have salt-adapted enzymes. In this case, the modifications needed to produce such salt resistant enzymes are of great interest to biotechnologists who are looking for enzymes which will operate in the harsh conditions of many industrial processes. The potential exploitation of *Dunaliella* will be discussed in some detail and other microalgal and bacterial systems will be discussed more briefly.

DUNALIELLA

This alga has been suggested as an ideal organism for commercial exploitation since three useful products can be extracted from it: glycerol, β-carotene and protein.[6] The high concentration of glycerol within high-salt-adapted cells (up to 80% of dry weight) has already been discussed. In addition, under high light conditions, some *Dunaliella* species produce β-carotene to counteract the harmful effects (photoinhibition) of excess light energy and β-carotene can make up 10–20% of the cell dry weight.[9,75] These very high levels of β-carotene are far above that needed for the avoidance of photoinhibition and it has been suggested that β-carotene acts a 'carbon sink'.[16] It has been observed that β-carotene is synthesized when intermediary metabolism is disrupted due to lack of substrates. However, photosynthesis must continue under these conditions to support glycerol synthesis and active ion transport systems, therefore it is suggested that the excess fixed carbon produced ends up as β-carotene, large amounts of which will not disrupt cell processes.[16] Thus, under conditions of high light intensity and high salinity, glycerol and β-carotene are synthesized in large amounts. Once these two compounds have been extracted, the remainder of the alga is largely protein which can be used as animal feed.[5]

USES OF GLYCEROL AND β-CAROTENE

Glycerol is used in drugs and cosmetics manufacture, in the food and beverage industries and in the production of urethane, cellophane and explosives.[5] Some glycerol is produced as a by-product of the fat and soap industry but the majority is derived from the petrochemical industry using propylene as the starting substance (Fig. 6.4), which means that the price of glycerol is dependent on the price of oil. β-Carotene is a valuable commodity being used as a colouring agent in foods such as margarine and cheese. In addition, a market is opening up for this product in the health food industry (β-carotene is a precursor of vitamin A) because many consumers now demand a natural product. β-carotene extracted from *Dunaliella* has approximately equal amounts of 9-*cis* and all-*trans* isomers, whereas synthetic β-carotene is 100% all-*trans* isomer. This was a potential problem since the all-*trans* form of β-carotene was considered to be a better source of vitamin A than the 9-*cis* isomer.[33] However, recent work has shown that this is not the case and that *Dunaliella* β-carotene is a good source of vitamin A in rats, chicks and humans.[8,10,67] β-Carotene may also be a cancer-preventative agent[59] and Schwartz *et al.*[118] succeeded in fusing protoplasts of the cyanobacterium *Spirulina* with *Dunaliella* protoplasts to form a hybrid organism, which has beneficial effects in treating oral cancer.

COMMERCIAL CULTURE SYSTEMS

To allow the production of glycerol and β-carotene, the *Dunaliella* cells must be supplied with high light intensities, high salinities and (to stimulate fast growth rates) high concentrations of CO_2. In hot countries such as Australia and Israel, with extremely arid regions, the first two of these requirements are naturally present and CO_2 is often available cheaply as stack gas (7–15% CO_2) from power plants.[23] Large vats in desert regions (or natural hypersaline lakes) have been inoculated with *Dunaliella* for the commercial production of glycerol and β-carotene. However, a major problem is that the highest concentration of β-carotene or glycerol per cell occurs under conditions where growth rate is sub-maximal. Chen and Chi,[23] investigating the outdoor production of glycerol by *Dunaliella*, proposed a two-stage process to overcome this problem. The first culture pond has a moderate salinity (about 1.5 M NaCl) which allows maximum growth to take place, the cells are then passed to a second culture pond which contains 4 M NaCl and this initiates massive glycerol accumulation by the cells. Such a set up could also be used for β-carotene production if the high salinity second pond was also low in nutrients. However, Borowitzka and Borowitzka[16] noted a number of drawbacks with this system:

1. Protozoa and other contaminants could grow in the relatively low salinity of the first culture pond.
2. β-Carotene producing *Dunaliella* tend to have higher salt optima than other *Dunaliella* sp. and thus can be outgrown in the first culture pond.
3. Increased labour is required and a relatively long time elapses before harvesting can take place.

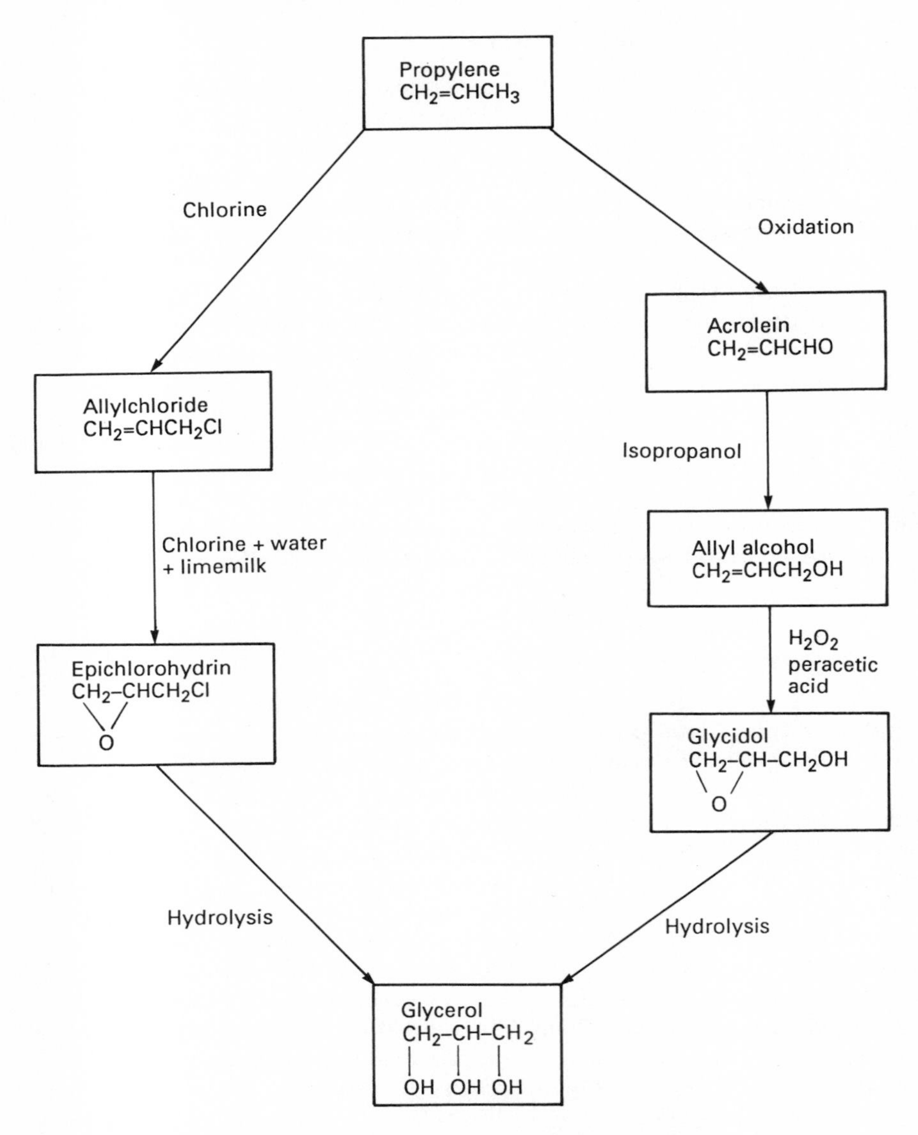

Fig. 6.4 The main methods of glycerol production used by the petrochemical industry (adapted from ref. 23).

The best approach seems to be a compromise whereby an intermediate salinity is used in a one-step process which although it is sub-optimal for growth and product synthesis, gives the best yield over a long period of time.[16]

HARVESTING

Finding a method of harvesting the algae which is efficient but not too expensive has been the major problem stopping widespread use of *Dunaliella* commercially. There are many potential methods described which include centrifugation, sedimentation, flotation and filtration.[107] Centrifugation is simply too expensive to use to separate the large quantity of biomass needed for a commercial operation. The most popular method is to flocculate the cells using inorganic salts such as aluminium or ferric chloride.[107] However, use of these compounds may be undesirable if the product is going to be consumed by animals and ultimately humans. Alternatively, positively charged polymers can be used as flocculants since they attract the negatively charged cells, and Mohn[99] suggests that potato starch and chitosans are suitable because they are recognized as non-toxic. After flocculation, the particles can either be allowed to settle or they can be induced to float using compressed air. The other major method is filtration, which suffers from rapid clogging of the filters. Nevertheless, as the design of filter apparatus improves, it may well become a viable alternative.[99]

ECONOMICS

At present there are companies in Australia, Britain and Israel using *Dunaliella* to produce glycerol; however, this is still on a relatively modest scale. Why? The answer lies in cost effectiveness. In 1981, Chen and Chi[23] believed that within a few years their process would compete financially with the petrochemical-based process. However, they projected a 15% annual increase in the price of crude oil into their calculation and thus predicted that in 1988 crude oil would cost $54 per barrel. At the time of writing (June 1988) the cost of crude oil is about $16–17 per barrel. Thus the production of glycerol from *Dunaliella* does not appear to be economically viable at present.

The situation with β-carotene is better, since consumers in the health food market appear to be willing to pay a premium for the natural product which is more expensive to produce than its synthetic competitor.[75] Nonetheless, more work is needed to optimize the cost-effectiveness of producing β-carotene from *Dunaliella*[16] and this is likely to include genetic engineering techniques, which will be discussed later.

Another way of making the production of β-carotene and glycerol economically viable is to couple it to other beneficial processes.[117] Williams *et al.*[130] suggested that a variety of chemicals could be synthesized from a crude extract of *Dunaliella* cells rich in glycerol, by using bacteria to ferment the glycerol to produce solvents and organic acids. The practicality of this all-microbial route to produce solvents was highly speculative, but Nakas *et al.*[104] succeeded in producing *n*-butanol, 1,3-propanediol and ethanol from *Dunaliella* biomass by allowing *Clostridium pas-*

teurianum to ferment the glycerol present. Although this process has not been scaled up to a commercial level, it illustrates one exciting possibility for the future.

OTHER MICROALGAE

Porphyridium (unicellular red alga) has a salt tolerance range of 0.25–1.0 M NaCl and shows biotechnological promise because it synthesizes the fatty acid arachidonic acid, which is an essential component of the human diet. *Porphyridium* also produces water-soluble extracellular polysaccharides which are of interest for their gelling properties, and a third possible product is the accessory photosynthetic pigment phycoerythrin which could be used as a non-toxic red pigment for food colouring.[125] Many marine diatoms and dinoflagellates (e.g. *Phaeodactylum*, *Navicula* and *Peridinium*) also synthesize arachidonic acid and the related fatty acid eicosapentaenoic acid; use of these algae as human dietary supplements is under investigation.[112]

Other compounds of commercial interest which can be extracted from algae are amino acids, pesticides, plant growth regulators, neurotoxins and pharmaceuticals.[14,59] Particular interest has been shown in using microalgae to produce lipids,[36] some species of the diatom *Chaetoceros* contain up to 80% of their dry weight as lipids when grown in high salinities.[15,68] Traditionally, lipids are used in the production of soaps, rubber and textiles. However, an on-going research programme at the Solar Energy Research Institute, Colorado, USA, is attempting to use algal lipids as the basis for the production of liquid fuels.[68]

SPIRULINA

Cyanobacteria belonging to the genus *Spirulina* are found in various habitats from brackish waters to saline lakes. High alkalinity is mandatory for good growth, and some of the densest populations of *Spirulina* are found in highly saline lakes of Africa and Central America which have high pH levels.[113] Indigenous human populations of these areas have used *Spirulina* as a food source for countless years. *Spirulina* is a good source of protein (50–70% of the dry weight) and in addition Richmond[113] lists the following effects that it has on humans:

1. Accelerates wound healing.
2. Photosynthetic accessory pigment, phycocyanin, may stimulate the immune system.
3. It is a good source of γ-linolenic acid, which is important in preventing degenerative diseases.

Spirulina can also contain large amounts of β-carotene whose potential uses have already been described.

HALOBACTERIUM

The biotechnological potential of extreme halophiles such as *Halobacterium* is much less direct than that just described for moderate halophiles or halotolerant microorganisms, since no industrially important compounds are synthesized in excess by *Halobacterium*. However, *Halobacterium* has enzymes which naturally work in high-salt conditions and these enzymes may be able to function and accelerate the desired reactions in the harsh conditions of some industrial processes which would immediately denature less resistant enzymes.

However, the genome of *Halobacterium* is very unstable (see next section), therefore, any role for *Halobacterium* in biotechnology is likely to depend on transferring some of its genetic information to another microbe (e.g. *E. coli*) using the techniques of genetic engineering, in which useful genes can hopefully be consistently over-expressed over long time periods on an industrial scale. This is beyond the capability of the techniques currently available.

Genetic aspects

An understanding of the molecular basis of salt tolerance could lead to the genetic manipulation of crop plants to make them capable of growth in more saline soils. This is a high priority aim, since much of the irrigation water used around the world is becoming more saline (e.g. Colorado River, USA) and salt tolerant crop plants are badly needed to maintain the food output of the world.[110] It was perceived that the higher plant system was too complex for the initial studies, but it was clear that the basic mechanism of salt tolerance was similar for bacteria (except for the halobacteria), algae and higher plants. The same compatible solutes (e.g. proline and betaine) are accumulated by both bacteria and higher plants when they are salt stressed. Therefore, what was needed was a bacterium which could successfully adapt to high salinities but whose genetics was well understood. The enteric bacteria such as *E. coli*, *Klebsiella* and *Salmonella* are ideal from the point of molecular genetics and they can grow in a range of salinities from 0 to about 1.0 M NaCl (Table 6.2). In the last few years much data have been published on the molecular basis of salt tolerance in enteric bacteria. Therefore, it seems appropriate to start this section with a discussion of the best understood systems in the enteric bacteria and then go on to discuss how much is known about the molecular basis of salt tolerance in *Halobacterium* and microalgae.

ENTERIC BACTERIA

Christian in 1955 was the first person to observe that the addition of proline to a *Salmonella* culture allowed the bacterium to grow in salinities which otherwise were completely inhibitory to growth.[24] More recently this observation was confirmed and extended to include other compatible solutes (e.g. betaine) and other enteric bacteria such as *Klebsiella* and *E. coli*.[91] Many enteric bacteria can tolerate salinities up to 0.5 M NaCl but addition of compatible solutes to the medium allows this salinity range to be extended to 1.0 M NaCl. These observations led to the development of selection procedures which could be used to isolate genes which controlled salt tolerance.[90] The first such gene (*proBA*) was isolated by Csonka in 1981 from a proline-overproducing mutant strain of *Salmonella typhimurium*.[28] Upon transfer to other *S. typhimurium* strains it conferred enhanced tolerance on the recipients (ref. 28 and Chapter 5 of this book). The biochemical mechanism of the increased salt tolerance occurs at the first step in proline biosynthesis catalysed by γ-glutamyl kinase; the enzyme in the mutant is 200-fold less sensitive to feedback inhibition than is the wild-type enzyme. This opened up the possibility of using plasmid cloning vectors to impart salt tolerance to other Gram-negative bacteria. Among the Gram-negative bacteria which could be manipulated in this way is the genus *Rhizobium*, whose members form symbiotic relationships with economically important plants such as soybean, alfalfa, beans and peas. This enables N_2-fixation to occur by the plant–*Rhizobium* symbiont. *Rhizobium* species are salt sensitive, thus clear benefits would accrue if their tolerance could be increased.[90]

The first attempts to construct a plasmid vector for Gram-negative bacteria using plasmids based on the narrow-host-range plasmid pBR322 failed, probably due to the inhibitory effect of a product from a gene cloned with *proBA*.[94] However, an alternative method using plasmids based on pQSR49 proved successful, although the stability of the recombinant plasmid in a variety of Gram-negative bacteria was dependent on the orientation of the insert and on the presence of foreign DNA on the plasmid.[66]

The proline uptake mechanism of *S. typhimurium* was also examined and it was found that, in addition to the two well-characterized proline permeases, there was a third permease (*proV*) which functions only in media of elevated salinity.[29] The relationship between proline transport and K^+ fluxes was also examined. In common with many other salt tolerant organisms, the initial reaction of enteric bacteria to elevated salinities is to accumulate K^+. The genetic control of two K^+ uptake systems (*Trk* and *Kdp*) has been elegantly demonstrated and the same group has shown that *proV* has a much higher affinity for betaine transport than for proline transport.[63] It appears that the elevated intracellular K^+ concentration is responsible for switching on the gene for betaine uptake (*proV*), and betaine replaces K^+ as the major compatible solute in *E. coli* and *S. typhimurium* cells at high salinities (>0.5 M). The osmoregulatory signal appears to depend on the intracellular ionic strength which is altered by the initial accumulation of K^+. Most recent evidence suggests that the elevated intracellular K^+ concentration affects the supercoiling of DNA and it is this which activates *proV* expression.

HALOBACTERIUM

Since the initial realization that halobacteria are archaebacteria and not eubacteria, much effort has been expended in studying their genome. The halobacteria have been used as representatives of the archaebacteria because they have proved easier to manipulate genetically than other archaebacteria such as the methanogens or thermoacidophiles.[35] This work led to various characteristics being assigned to archaebacteria concerning the structure and function of their genome.[32] One major characteristic of *Halobacterium* is the instability of the genome. Spontaneous mutations affecting production of gas vacuoles or the pigment bacterioruberin occur at frequencies as high as one in 10^2 cells plated.

Another feature of *Halobacterium* is the high level of extrachromosomal or satellite DNA. This can account for up to 36% of the total DNA in some species and it appears to consist of plasmids with reiterated sequences. This may be the key to the high genetic variability, since some regions of sequence homology may promote recombination between chromosomal and plasmid DNA.[79]

Halobacterium would appear to be an almost infinitely variable genetic system and it is tempting to suggest that such a system is ideal for applying genetic engineering techniques, but the genetic variability could cause severe problems in isolating unique genes and frustrate attempts at genetic mapping. In addition, no transformation system is as yet available for the introduction and propagation of recombinant DNA in *Halobacterium*.[32] One final puzzle, pointed out by Kushner,[79] is that, despite the large genetic variability found in the laboratory, only moderate variation appears to occur in natural populations of halobacteria. At the moment there is no good explanation as to why this should be so, although it has been suggested that only a small percentage of *Halobacterium* species present in natural habitats can be isolated using conventional high salt media.[127]

MICROALGAE

The genetics of algae has been intensively studied with the majority of work being done using the unicellular green alga *Chlamydomonas reinhardtii*.[114] Classical genetic techniques have been used on spontaneous mutants or on mutants produced by ultraviolet light or chemical treatment to map the genes involved. *C. reinhardtii* has two mating types + and − and, after sexual reproduction, a tetrad of four daughter cells is formed. Many of the cell characters are determined by a single pair of alleles, are segregated in a 2:2 ratio, and result from single gene mutations. Many interesting mutants of *C. reinhardtii* have been studied, particularly those which are deficient in some part of the photosynthetic apparatus (e.g. refs 89 and 97). This allows the genetic control of photosynthesis to be examined. The genetics of *Chlamydomonas* is well understood, and a system exists for transforming cells by the uptake of naked DNA into the cell, but it is not without its problems.[26] To improve the efficiency of transformation, Craig *et al.*[26] have suggested using 'minichromosomes' consisting of a bacterial plasmid, an eukaryotic biosynthetic gene, an eukaryotic chromosomal replicator sequence and a centromere sequence. The biosynthetic gene would be one in which the host was defective and it would thus act as a marker for the success of the transformation. The replicator and centromere make it more likely that the 'minichromosome' would be replicated in the host *Chlamydomonas* cell.[26]

C. reinhardtii is a freshwater alga but it is closely related phylogenetically to *Dunaliella*. Perhaps similar techniques can be used to probe the genetic control of salt tolerance in *Dunaliella*. Mutants of *Dunaliella* can be produced by treatment with chemical mutagens (e.g. methyl nitrosoguanidine) and Brown *et al.*[21] succeeded in isolating a mutant deficient in carbonic anhydrase activity which was salt sensitive. An alternative approach has been used by Lee and Tan[88] to achieve genetic recombination between two algae, *Porphyridium* and *Dunaliella*, which are phylogenetically dissimilar. True protoplasts of both algae were produced by treatment with pectinase and cellulysin, and they were induced to fuse by polyethylene glycol treatment. Hybrid organisms were produced which were essentially *Porphyridium* but with a much increased salt tolerance, and other hybrids were *Dunaliella*-like but had increased resistance to a range of antibiotics — a feature of *Porphyridium*. Thus there appears little doubt that genetic recombination had occurred.[88]

If stable mutants which are salt sensitive can be isolated from *Dunaliella* and if a transformation system can be developed to add DNA back into the mutant cells, then complementation can be used to identify genes involved in salt tolerance. Given the present rapid rate of improvement in genetic techniques, this is by no means an impossible goal to aim for. It is likely that the construction of 'minichromosomes' described for *Chlamydomonas* could find an application with *Dunaliella*. Classical genetic techniques may be of less value, since sexual reproduction in *Dunaliella* is poorly understood and a great deal of time and effort would be needed to bring the level of understanding of *Dunaliella* genetics up to that of *Chlamydomonas* genetics.

Future prospects

Although work remains to be done to fully characterize the physiological mechanisms of halotolerance, the most exciting developments are likely to come in the commercial utilization of halophiles and in the unravelling of the genetic control of salt tolerance. The use of halotolerant algae in mass outdoor cultures is by no means a new idea, and pilot plants and full commercial operations have been set up for the production of glycerol and β-carotene from *Dunaliella*.[5,23] The feasibility of this system has been proved and only the relatively cheap price of oil stops it from being more widely used. Nevertheless, fossil fuel deposits have a limited lifespan and it would be wise for efforts to be made to find more renewable sources of useful chemicals like the *Dunaliella* system. Ten years ago this type of research was being actively pursued in the wake of the energy crisis in the mid-seventies. However, since this crisis eased, less money has been invested in biomass production using renewable energy. This is clearly short-sighted and it is to be hoped that such research will be fully supported in the coming years.

Many of the most exciting biotechnological possibilities depend on an extensive knowledge of the genetic control of salt tolerance and, in enteric bacteria such as *E. coli*, this depth of understanding already exists.[63] Since these microbes use similar compatible solutes to salt resistant higher plants, the possibility of cloning bacterial genes into crop plants to make them more salt resistant is a very exciting one. As yet, the techniques required to do this by direct means do not exist but it is likely that they will be developed in the near future.[93] The importance of this advance cannot be overestimated, since increasing salination of irrigation water is a worldwide problem.

A second approach to this problem would be to examine the genetic control of salt tolerance in *Dunaliella*, since this alga is the most salt tolerant eukaryote known and it has a photosynthetic mechanism identical to many higher plants. It would be necessary to understand the genetic control of salt tolerance in *Dunaliella* as well as it is currently understood in *E. coli*. This will involve much fundamental research before it can be applied in a practical way to improving crop yields in saline soils.

The role of *Halobacterium* species in biotechnology is likely to be a minor one because their physiological adaptation mechanism to elevated salinities is so different from all other halophiles. They may have a role to play in helping produce unusual enzymes for industrial processes but this role will probably be taken up largely by thermophilic microbes, since enzymes which are active at high temperatures are usually the prime requirement for many industrial processes (see Chapter 1).

Conclusions

An attempt has been made to explain the mechanism of salt tolerance or adaptation in halotolerant and halophilic microorganisms taking *Dunaliella* and *Halobacterium* as the main examples. These two organisms were chosen as representatives of eukaryote and prokaryote halophiles, respectively, and they are adapted to life in highly saline environments by fundamentally different mechanisms. *Dunaliella* produces a large amount of an organic compatible solute and thereby maintains low cytoplasmic salt concentrations. *Halobacterium* produces no organic compatible solutes but tolerates high internal salt concentrations by having highly salt-resistant enzymes. The potential commercial exploitation of halotolerant/halophilic microorganisms and the closely related question of the genetic control of salt tolerance were both discussed. It is clear that exciting possibilities exist with halophiles to increase our understanding of the fundamental nature of salt tolerance and in doing so open the way for the commercial exploitation of these organisms as significant biomass producers of the next century.

Acknowledgements

I would like to thank J.R. Blackwell and D.J. Kelly for their critical reading of parts of the manuscript.

References

1. Adams, R., Bygraves, J., Kogut, M. and Russell, N.J. (1987) *J. Gen. Microbiol.*, **133**, 1861–70.
2. Belmans, D. and Van Laere, A. (1987) *Plant Cell Environ.*, **10**, 185–90.
3. Belmans, D. and Van Laere, A. (1988) *Arch. Microbiol.*, **150**, 109–12.
4. Ben-Amotz, A. and Avron, M. (1973) *Plant Physiol.*, **51**, 875–8.
5. Ben-Amotz, A. and Avron, M. (1980), in G. Shelef and C.J. Soeder (eds) *Algae Biomass*, Amsterdam, Elsevier, pp. 603–10.

6. Ben-Amotz, A. and Avron, M. (1981) *Trends Biochem. Sci.*, **6**, 297–9.
7. Ben-Amotz, A. and Avron, M. (1983) *Annu. Rev. Microbiol.*, **37**, 95–119.
8. Ben-Amotz, A., Edelstein, S. and Avron, M. (1986) *British Poultry Science*, **27**, 613–19.
9. Ben-Amotz, A., Lers, A. and Avron, M. (1988) *Plant Physiol.*, **86**, 1286–91.
10. Ben-Amotz, A., Mokady, S. and Avron, M. (1988) *Brit. J. Nutrition*, **59**, 443–9.
11. Borowitzka, L.J. (1981), in L.G. Paleg and D. Aspinall (eds) *Physiology and Biochemistry of Drought Resistance in Plants*, London, Academic Press, pp. 97–130.
12. Borowitzka, L.J. and Brown, A.D. (1974) *Arch. Microbiol.*, **96**, 37–52.
13. Borowitzka, L.J., Kessly, D.S. and Brown, A.D. (1977) *Arch. Microbiol.*, **113**, 131–8.
14. Borowitzka, M.A. (1988), in L.J. Borowitzka and M.A. Borowitzka (eds) *Micro-algal Biotechnology*, Cambridge University Press, pp. 154–96.
15. Borowitzka, M.A. (1988), in L.J. Borowitzka and M.A. Borowitzka (eds) *Micro-algal Biotechnology*, Cambridge University Press, 257–87.
16. Borowitzka, M.A. and Borowitzka, L.J. (1988), in L.J. Borowitzka and M.A. Borowitzka (eds) *Micro-algal Biotechnology*, Cambridge University Press, pp. 27–58.
17. Brock, T.D. (1969) *Symp. Soc. Gen. Microbiol.*, **19**, 15–42.
18. Brown, A.D. (1964) *Bacteriol Rev.*, **28**, 296–329.
19. Brown, A.D. (1976) *Bacteriol Rev.*, **40**, 803–46.
20. Brown, A.D. (1983), in O.L. Lange, P.S. Nobel and C.B. Osmond (eds) *Encyclopedia of Plant Physiology, 12c*, New York, Springer-Verlag, pp. 137–62.
21. Brown, A.D., Goyal, A., Larsen, H. and Lilley, R.McC. (1987) *Arch. Microbiol.*, **147**, 309–14.
22. Brown, F.F., Sussman, I., Avron, M. and Degani, H. (1982) *Biochim. Biophys. Acta*, **690**, 165–73.
23. Chen, B.J. and Chi, C.H. (1981) *Biotech. Bioeng.*, **23**, 1267–88.
24. Christian, J.H.B. (1955) *Aust. J. Biol. Sci.*, **8**, 490–7.
25. Christian, J.H.B. and Waltho, J. (1962) *Biochim. Biophys. Acta,* **65**, 506–8.
26. Craig, R., Reichelt, B.Y. and Reichelt, J.L. (1988), in L.J. Borowitzka and M.A. Borowitzka (eds) *Micro-algal Biotechnology*, Cambridge University Press, pp. 415–55.
27. Craigie, J.S. and McLachlan, J. (1964) *Can. J. Bot.*, **42**, 777–8.
28. Csonka, L.N. (1981) *Mol. Gen. Genet.*, **182**, 82–6.
29. Csonka, L.N. (1982) *J. Bacteriol.*, **151**, 1433–43.
30. Curtain, C.C., Looney, F.D., Regan, D.L. and Ivancic, N.M. (1983) *Biochem. J.*, **213**, 313–6.
31. Degani, H., Sussman, I., Peschek, G.A. and Avron, M. (1985) *Biochim. Biophys. Acta*, **846**, 313–23.
32. Dennis, P.P. (1986) *J. Bacteriol.*, **168**, 471–8.
33. Deuel, H.J., Johnson, C., Sumner, E., Polgar, A. and Zechmeister, L. (1944) *Arch. Biochem.*, **5**, 107–12.
34. Dickson, D.M.J. and Kirst, G.O. (1986) *Planta*, **167**, 536–43.
35. Doolittle, W.F. (1985), in C.R. Woese and R.S. Wolfe (eds) *The Bacteria Vol. VIII Archaebacteria*, London, Academic Press, pp. 545–60.
36. Dubinsky, Z., Berner, T. and Aaronson, S. (1978) *Biotech. Bioeng. Symp.*, **8**, 51–68.
37. Duschl, A. and Wagner, G. (1986) *J. Bacteriol.*, **168**, 548–52.
38. Eisenberg, H. and Wachtel, E.J. (1987) *Annu. Rev. Biophys. Biophys. Chem.*, **16**, 69–92.
39. Enhuber, G. and Gimmler, H. (1980) *J. Phycol.*, **16**, 524–32.
40. Galinsky, E. and Truper, H.G. (1982) *FEMS Lett.*, **13**, 357–60.
41. Gilmour, D.J., Hipkins, M.F. and Boney, A.D. (1984) *J. Exp. Bot.*, **35**, 18–27.
42. Gilmour, D.J., Hipkins, M.F. and Boney, A.D. (1984) *J. Exp. Bot.*, **35**, 28–35.
43. Gilmour, D.J., Kaaden, R. and Gimmler, H. (1985) *J. Plant Physiol.*, **118**, 111–26.

44. Gimmler, H. and Lotter, G. (1982) *Z. Naturforsch.*, **37c**, 1107–14.
45. Gimmler, H. and Moller, E.-M. (1981) *Plant Cell Environ.*, **4**, 367–75.
46. Gimmler, H., Kaaden, R., Kirchner, U. and Weyand, A. (1984) *Z. Pflanzenphysiol.*, **114**, 131–50.
47. Ginzburg, M. (1969) *Biochim. Biophys. Acta*, **173**, 370–6.
48. Ginzburg, M. (1981) *J. Exp. Bot.*, **32**, 333–40.
49. Ginzburg, M. and Ginzburg, B.Z. (1981) *Brit. Phycol. J.*, **16**, 313–24.
50. Ginzburg, M. and Ginzburg, B.Z. (1985) *J. Exp. Bot.*, **36**, 701–12.
51. Ginzburg, M., Sachs, L. and Ginzburg, B.Z. (1971) *J. Membr. Biol.*, **5**, 78–101.
52. Goyal, A., Lilley, R.McC. and Brown, A.D. (1986) *Plant Cell Environ.*, **9**, 703–6.
53. Goyal, A., Brown, A.D. and Gimmler, H. (1987) *J. Plant Physiol.*, **127**, 77–96.
54. Grant, W.D. and Ross, H.N.M. (1986) *FEMS Microbiol. Rev.*, **39**, 9–15.
55. Griffin, D.M. and Luard, E.J. (1979), in M. Shilo (ed.) *Strategies of Microbial Life in Extreme Environments*, Life Sciences Reports Vol. 13, Weinheim, Verlag-Chemie, pp. 49–63.
56. Hajibagheri, M.A., Gilmour, D.J., Collins, J.C. and Flowers, T.J. (1986) *J. Exp. Bot.*, **37**, 1725–32.
57. Hanna, K., Bengis-Garber, C., Kushner, D.J., Kogut, M. and Kates, M. (1984) *Can. J. Microbiol.*, **30**, 669–75.
58. Harel, M., Shoham, M., Frolow, F., Eisenberg, H., Mevarech, M., Yonath, A. and Sussman, J.L. (1988) *J. Mol. Biol.*, **200**, 609–10.
59. Harvey, W. (1988) *Bio/technology*, **6**, 487–92.
60. Haus, M. and Wegmann, K. (1984) *Physiol. Plant.*, **60**, 283–8.
61. Hellebust, J.A. (1976) *Annu. Rev. Plant Physiol.*, **27**, 485–505.
62. Hellebust, J.A. (1985) *Plant and Soil*, **89**, 69–81.
63. Higgins, C.F., Cairney, J., Stirling, D.A., Sutherland, L. and Booth, I.R. (1987) *Trends in Biochem. Sci.*, **12**, 339–44.
64. Hoshaw, R.W. and Maluf, L.Y. (1981) *Phycologia*, **20**, 199–206.
65. Imhoff, J.F. and Rodriguez-Valera, F. (1984) *J. Bacteriol.*, **160**, 478–9.
66. Jakowec, M.W., Tombras Smith, L. and Dandekar, A.M. (1985) *Appl. Environ. Microbiol.*, **50**, 441–6.
67. Jensen, C.D., Howes, T.W., Spiller, G.A., Pattison, T.S., Wittam, J.H. and Scala, J. (1987) *Nutrition Reports International*, **35**, 413–22.
68. Johnson, D.A. and Sprague, S. (1987) Annual Report, Solar Energy Research Institute, Golden, Colorado, USA.
69. Katz, A., Kabach, H.R. and Avron, M. (1986) *FEBS Lett.*, **202**, 141–4.
70. Ken-Dror, S., Preger, R. and Avi-Dor, Y. (1986) *Arch. Biochem. Biophys.*, **244**, 122–7.
71. Ken-Dror, S., Preger, R. and Avi-Dor, Y. (1986) *FEMS Microbiol. Rev.*, **39**, 115–20.
72. Kirst, G.O. (1977) *Oecologia*, **28**, 177–89.
73. Kirst, G.O. (1979) *Ber. Deut. Bot. Ges.*, **92**, 31–42.
74. Kirst, G.O. (1980) *Z. Pflanzenphysiol.*, **98**, 35–42.
75. Klausner, A. (1986) *Bio/technology*, **4**, 947–53.
76. Konishi, T. and Murakami, N. (1988) *FEBS Lett.*, **226**, 270–4.
77. Kushner, D.J. (1968) *Adv. Appl. Microbiol.*, **10**, 73–99.
78. Kushner, D.J. (1978), in D.J. Kushner (ed.) *Microbial Life in Extreme Environments*, London, Academic Press, pp. 318–68.
79. Kushner, D.J. (1985), in C.R. Woese and R.S. Wolfe (eds) *The Bacteria Vol. VIII Archaebacteria*, London, Academic Press, pp. 171–214.
80. Kushner, D.J. (1986) *FEMS Microbiol. Rev.*, **39**, 121–7.
81. Kushner, D.J. (1988) *Can. J. Microbiol.*, **34**, 482–6.

82. Lanyi, J.K. (1974) *Bacteriol. Rev.*, **38**, 272–90.
83. Larsen, H. (1973) *Antonie van Leeuwenhoek*, **39**, 383–96.
84. Larsen, H. (1980), in A. Nissenbaum (ed.) *Hypersaline Brines and Evaporitic Environments, Developments in Sedimentology Vol. 28*, Amsterdam, Elsevier, pp. 23–39.
85. Larsen, H. (1981), in M.P. Starr, H. Stolp, H.G. Truper, A. Balows and H.G. Schlegel (eds) *The Prokaryotes, a Handbook on Habitats, Isolation and Identification of Bacteria*, New York, Springer-Verlag, pp. 985–94.
86. Larsen, H. (1986) *FEMS Microbiol. Rev.*, **39**, 3–7.
87. Latorella, A.H. and Vadas, R.L. (1973) *J. Phycol.*, **9**, 273–7.
88. Lee, Y.K. and Tan, H.-M. (1988) *J. Gen. Microbiol.*, **134**, 635–41.
89. Lemaire, C., Girard-Bascou, J. and Wollman, F.-A. (1987), in J. Biggens (ed.) *Progress in Photosynthesis Research, Vol. IV*, Dordrecht, Martinus Nijhoff, pp. 655–8.
90. Le Rudulier, D. and Valentine, R.C. (1982) *Trends in Biochem. Sci.*, **7**, 431–3.
91. Le Rudulier, D., Strom, A.R., Dandekar, A.M., Smith, L.T. and Valentine, R.C. (1984) *Science*, **224**, 1064–8.
92. Maeda, M. and Thompson, G.A. (1986) *J. Cell Biol.*, **102**, 289–97.
93. Magnien, E. and de Nettancourt, D. (1985) *Genetic Engineering of Plants and Microorganisms Important for Agriculture*, Dordrecht, Martinus Nijhoff.
94. Mahan, M.J. and Csonka, L.N. (1983) *J. Bacteriol.*, **156**, 1249–62.
95. Marengo, T., Lilley, R. McC. and Brown, A.D. (1985) *Arch. Microbiol.*, **142**, 262–8.
96. McLusky, D.S. (1981) *The Estuarine Ecosystem*, London, Blackie.
97. Merchant, S. and Bogorad, L. (1987), in J. Biggins (ed.) *Progress in Photosynthetic Research, Vol. IV*, Dordrecht, Martinus Nijhoff, pp. 663–6.
98. Miller, K.J. (1985) *J. Bacteriol.*, **162**, 263–70.
99. Mohn, F.H. (1988), in L.J. Borowitzka and M.A. Borowitzka (eds) *Micro-algal Biotechnology*, Cambridge University Press, pp. 395–414.
100. Morgan, H., Ginzburg, M. and Ginzburg, B.Z. (1987) *Biochim. Biophys. Acta*, **924**, 54–66.
101. Munns, R., Greenway, H. and Kirst, G.O. (1983), in O.L. Lange, P.S. Nobel and C.B. Osmond (eds) *Encyclopedia of Plant Physiology, 12c*, New York, Springer-Verlag, pp. 99–135.
102. Murakami, N. and Konishi, T. (1988) *Biochimie*, **70**, 819–26.
103. Murakami, N. and Konishi, T. (1988) *J. Biochem.*, **103**, 231–6.
104. Nakas, J.P., Schaedle, M., Parkinson, C.M., Cooney, C.E. and Tannenbaum, S.W. (1983) *Appl. Environ. Microbiol.*, **46**, 1017–23.
105. Nissenbaum, A. (1975) *Microbial Ecol.*, **2**, 139–61.
106. Oren, A. (1988) *Antonie van Leeuwenhoek*, **54**, 267–77.
107. Oswald, W.J. (1988), in L.J. Borowitzka and M.A. Borowitzka (eds) *Micro-algal Biotechnology*, Cambridge University Press, pp. 357–94.
108. Pick, U., Karni, L. and Avron, M. (1986) *Plant Physiol.*, **81**, 92–6.
109. Post, F.J. (1977) *Microbial Ecol.*, **3**, 143–65.
110. Rains, D.W., Valentine, R.C. and Hollaender, A. (1980) *Genetic Engineering of Osmoregulation. Impact on Plant Productivity for Food, Chemicals and Energy*, London, Plenum Press.
111. Reed, R.H. (1986) in R.A. Herbert and G.A. Codd (eds) *Microbes in Extreme Environments*, London, Academic Press, pp. 55–81.
112. Regan, D.L. (1988), in L.J. Borowitzka and M.A. Borowitzka (eds) *Micro-algal Biotechnology*, Cambridge University Press, pp. 137–50.
113. Richmond, A. (1988), in L.J. Borowitzka and M.A. Borowitzka (eds) *Micro-algal Biotechnology*, Cambridge University Press, pp. 85–121.

114. Rochaix, J.-D. and van Dillewijn, J. (1982) *Nature*, **296**, 70–2.
115. Ross, H.N.M. and Grant, W.D. (1985) *J. Gen. Microbiol.*, **131**, 165–73.
116. Russell, N.J., Adams, R., Bygraves, J. and Kogut, M. (1986) *FEMS Microbiol. Rev.*, **39**, 103–07.
117. Sanderson, J.E., Wise, D.L. and Augenstein, D.C. (1978) *Biotech. Bioeng. Symp.*, **8**, 131–51.
118. Schwartz, J., Shklar, G., Reid, S. and Trickler, D. (1988) *Nutrition and Cancer*, **11**, 127–37.
119. Sheffer, M., Fried, A., Gottlieb, H.E., Tietz, A. and Avron, M. (1986) *Biochim. Biophys. Acta*, **857**, 165–72.
120. Stewart, W.D.P. (1983) *Soc. Gen. Microbiol. Symp.*, **34**, 1–35.
121. Stoeckenius, W. and Bogomolni, R.A. (1982) *Annu. Rev. Biochem.*, **52**, 587–616.
122. Tokuda, H. (1986) *Methods in Enzymol.*, **125**, 520–30.
123. Udagawa, T., Unemoto, T. and Tokuda, H. (1986) *J. Biol. Chem.*, **261**, 2616–22.
124. Vidal, M.C. and Cazzulo, J.J. (1976) *Experientia*, **32**, 441–2.
125. Vonshak, A. (1988), in L.J. Borowitzka and M.A. Borowitzka (eds) *Micro-algal Biotechnology*, Cambridge University Press, pp. 122–34.
126. Vreeland, R.H. (1987) *CRC Crit. Rev. in Microbiol.*, **14**, 311–56.
127. Wais, A.C. (1988) *FEMS Microbiol. Ecol.*, **53**, 211–16.
128. Wegmann, K. (1971) *Biochim. Biophys. Acta*, **234**, 317–23.
129. Werber, M.M., Sussman, J.L. and Eisenberg, H. (1986) *FEMS Microbiol. Rev.*, **39**, 129–35.
130. Williams, L.A., Foo, E.L., Foo, A.S., Kuhn, I. and Heden, C.-G. (1978) *Biotech. Bioeng. Symp.*, **8**, 115–30.
131. Woese, C.R. (1981) *Scientific American*, **244**, 94–106.
132. Yancey, P.H., Clark, M.E., Hand, S.C., Bowlus, R.D. and Somero, G.N. (1982) *Science*, **217**, 1214–22.
133. Zidan, M.A., Hipkins, M.F. and Boney, A.D. (1987) *J. Plant Physiol.*, **127**, 461–9.

CHAPTER 7

Metal tolerance

G.M. Gadd

Introduction

There are approximately sixty-five elements that exhibit metallic properties which may be termed 'heavy metals'.[34] However, although this is a familiar and widely used term, it is clearly imprecise. In biological contexts, the principal chemical species are cations and the many definitions based on, for example, density, reactivity or position in the periodic table include a wide variety of elements with diverse chemical and biological properties.[45,47] In this chapter, the term 'heavy metal' will be used in its widest sense and, as well as the more commonly encountered elements such as mercury, lead, copper, cadmium, zinc, manganese, nickel and cobalt, attention will also be paid to rarer elements like gold and silver, organometallic compounds and radionuclides such as uranium and thorium.

A general feature of the heavy metals is their well-known potential toxicity towards microbial and other life forms which is the basis of many biocidal preparations. However, many 'trace metals' are essential for growth and metabolism at low concentrations, e.g. Cu, Zn, Fe, Ni, Mn, Co, and microorganisms possess mechanisms of varying specificity for their intracellular accumulation from the external environment. In contrast, many other metals have no essential biological functions, e.g. Pb, Sn, Cd, Al, Hg, but still may be accumulated.[30,34,35,47,99,118]

The evolutionary selection of essential elements may partly reflect their solubility in water under anaerobic conditions. Thus, biologically essential iron is present as water soluble Fe(II) salts under anaerobic conditions and would be transportable by primitive anaerobic bacteria. However, under similar conditions, aluminium, lead and tin are insoluble unless the pH is very low.[130] In addition, non-essential metals are generally of low abundance in the biosphere and should therefore not compete with specific transport systems for essential elements.[130] However, due to industrial activities and the resulting changes in global elemental distributions and pollution of aquatic and terrestrial habitats, microorganisms are increasingly exposed to potentially toxic conditions and therefore may need to respond using a variety of strategies that ensure survival and reproduction.[45,47,130]

Some metal–microbe interactions, where accumulation by the biomass leads to removal from the external environment, are of current biotechnological interest for both the recovery of valuable metals and the detoxification of polluted effluents.[121] An understanding of microbial responses towards heavy metals is also relevant to the preservation and protection of natural and synthetic materials since inorganic and organometallic compounds are still widely used as microbiocidal agents. This can result in further pollution of terrestrial and aquatic habitats by run-off and leaching.

Habitats

In natural environments, average abundances of heavy metals are generally low and much of that sequestered in sediments, soils and mineral deposits may be biologically unavailable. Although elevated metal levels can occur in rather specific natural locations, e.g. deep-sea vents, hot springs and volcanic soils, it is mainly as a result of industrial activities that ecosystems are increasingly subject to heavy metal pollution (Table 7.1).

A wide variety of industrial activities have accelerated the mobilization of many heavy metals above rates of natural geochemical cycling and there is increased deposition in aquatic and terrestrial environments as well as release into the atmosphere. Atmospheric emissions of Cd, Zn and Pb may result from processes such as mining and ore processing as well as the burning of fossil fuels[7] whereas elemental Hg and organomercurials enter the atmosphere from soils, rocks and sewage works largely as a result of microbial activity.[111] Changing agricultural practices, e.g. more efficient ploughing and the use of fertilizers, which accelerate natural rates of biological activity in soil, can indirectly lead to increased formation of volatile mercury compounds and in the natural flux of Hg, this 'degassing' is of dominating importance.[67] Atmospheric emissions eventually enter terrestrial and aquatic environments.

Oceans are a large reservoir of heavy metals and radionuclides although, with dilution and sequestration in sediments, average concentrations remain low and toxic effects may not be readily observed except at local sites of pollution. For example, human activities enter $2\text{–}7 \times 10^4$ tons Hg year^{-1} to the biosphere from industrial activities and the burning of fossil fuels. Over the last hundred years, such levels are relatively negligible and only approximate to $<1\%$ of the total amount of Hg in the oceans. However, on a local scale, pollution from industrial sources can result in toxic effects that may be intractable and extreme in many cases.[65,111]

Heavy metals exert their harmful effects in many ways, although all the major

Table 7.1 Natural and anthropogenic sources of heavy metals, radionuclides and organometallic compounds in the environment. This selection is not exhaustive and it should be realized that almost every industrial activity can lead to altered and/or increased metal distribution in natural habitats

Weathering processes	Chlorine–alkali plants
Deep-sea vents	Battery manufacture
Volcanic activities	Vehicular emissions
Microbial solubilization	Nuclear waste
Metal transformations	Sewage treatment effluents
Atmospheric deposition	Pigment manufacture
Mining and ore processing	Seed dressings
Iron and steel production	Animal feed additives
Metal plating	Brewery and distillery wastes
Combustion of fossil fuels	Domestic waste
Mine waters and leachates	Agricultural fungicides
Industrial effluents and discharges	Wood and stone preservatives
Industrial accidents	Biocides and disinfectants
Fertilizer manufacture	Pharmaceuticals
Sewage sludge fertilizers	Industrial catalysts and stabilizers
Photographic waste	Antifouling paint biocides

mechanisms of toxicity are a consequence of the strong coordinating abilities of metal ions.[92] For toxicity to be manifest, metals must enter microbial cells and, as will be discussed, transport systems of varying specificity may be utilized.[45] In view of the wide variety of ligands found in living cells[88] and the range of potentially toxic interactions that can occur, almost every index of microbial metabolism and activity in aquatic or terrestrial ecosystems can be dramatically reduced under conditions of metal pollution. These include primary productivity, nitrogen fixation, biogeochemical cycling of C, N, S, P and other elements, litter decomposition, enzyme synthesis and activity in soils, sediments and waters.[6] Because of the microbial involvement in such ecological processes, as well as in plant growth and productivity and symbiotic associations, it follows that heavy metal pollution can have serious and long-ranging effects, ultimately posing a threat to humans by bioaccumulation and transfer through food chains.[6,34,47]

It should be stressed that in a polluted environment, heavy metals may not be the only hazardous components encountered by the microflora. Other toxicants may be present and, in addition, environments spoilt by industrial disturbance are frequently nutrient poor and of extreme pH, e.g. mine spoils and leachates. Aquatic environments vary in organic matter content and exhibit a range of salinities. In a more general sense, the physico-chemical characteristics of a given ecosystem into which metals are deposited determine the form and biological availability of heavy metals and therefore their toxicity.[5–7,47] Such abiotic factors may include pH, oxidation–reduction potential (E_h), aeration, inorganic anions and cations, particulate and soluble organic matter, clay minerals, hydrous metal oxides, salinity, temperature and hydrostatic pressure.[6] Where such factors decrease biologically available concentrations of metals, toxicity is often reduced. Binding interactions with organic environmental components, e.g. humic and fulvic acids, proteins, organic chelating agents, can remove metals from solution or form complexes or chelates which may not be as readily accumulated by microorganisms as free cations. A decreased pH may increase metal availability, whereas towards and above neutrality, insoluble oxides, hydroxides and carbonates may form. In addition, the pH may affect other aspects of cell physiology, e.g. nutrient transport and also the rate and extent of complexation to organic and inorganic components. Thus, pH effects may be complex and, depending on the particular system examined, toxicity may increase, decrease or be unaffected.[6,47] However, in many examples, toxicity is reduced at acidic pH values, despite the often increased metal availability, and several extreme examples of microbial metal tolerance are confined to acidic conditions.[47] Anions such as CO_3^{2-}, S^{2-}, PO_4^{3-}, etc. can reduce metal availability by precipitation of corresponding carbonates, sulphides and phosphates.

The organisms

In contrast to those 'extreme' environments characterized by a relatively specific, restricted microflora, a wide range of organisms from all the major groups may be found in metal-polluted habitats and the ability to survive and grow in the presence of potentially toxic metal concentrations is frequently encountered. Thus, generalizations about a so-called 'specific microflora' of polluted habitats cannot be made with any certainty. In general terms, heavy metals are believed to affect microbial populations in nature by reducing abundance and species diversity and selecting for a resistant population.[6,34] However, these general assumptions require critical appraisal.

The effect of heavy metals on microbial abundance in natural habitats varies depending on the metals and microorganisms and the physico-chemical attributes of that environment.[34,47] General reductions in the numbers of bacteria, including actinomycetes, and fungi have often been recorded in soils polluted with Cu, Cd, Pb, As and Zn.[6] Along a steep gradient of Cu and Zn in soil towards a brass mill, fungal biomass decreased by about 75%[91] whereas the total number of fungi in a glucose-supplemented soil was reduced by the addition of Cd or Zn, the former metal having the greatest toxic effect.[15] Similar reductions in numbers of aquatic cyanobacteria and algae in response to metal pollution have also been noted.[96] However, numerical estimates, if viewed in isolation, provide no indication as to whether the microbial populations have changed in any way as a result of metal exposure. For example, a Zn- and Cd-polluted soil located near a smelter had lower numbers of actinomycetes and fungi but not total bacteria as compared with unpolluted control soil. However, the latter contained greater numbers of *Nitrosomonas* sp.[6] Combinations of Cd and Zn reduced the numbers of bacteria and actinomycetes to a greater extent than numbers of fungi in a glucose-supplemented soil.[16] In a study where Cu was added to a simulated marine ecosystem, an increase in heterotrophic bacteria occurred which was thought to be the result of bacteria which survived the Cu exposure using the organic substrates released from Cu-sensitive organisms.[122]

Thus, numerical estimates may provide little meaningful information without consideration of changes in microbial groups and species and it seems evident that such changes can occur in response to metal exposure. As in other so-called extreme environments, there can be a reduction in species diversity.[6,34,82] In relation to pollution of leaf surfaces by aerially deposited Pb, Cd or Zn, bacteria appeared more sensitive than fungi, and on polluted oak leaves *Aureobasidium pullulans* and *Cladosporium* sp. were the most numerous organisms.[14] In fact, numbers of *A. pullulans* showed a good positive correlation with lead concentrations, whether derived from industrial or vehicular sources and it often became the dominant organism, in some cases comprising up to 97% of the leaf surface population on a numerical basis.[14,82] In contrast, the ballistospore-producing yeast *Sporobolomyces roseus* and heterotrophic bacteria were very low or absent from polluted samples and numbers of *S. roseus* showed a significant negative correlation with increasing lead concentrations.[82] In Cu- and Zn-polluted soil, the fungi *Geomyces* and *Paecilomyces* sp. and some sterile forms increased with increasing pollution, whereas *Penicillium* and *Oidiodendron* sp. declined at polluted sites.[91] *Trichocladium asperum*, *Trichoderma hamatum*, *Zygorrhynchus moelleri* and *Chrysosporium pannorum* were isolated more frequently from an organomercurial-treated golf green than from untreated locations, whereas *Chaetomium*, *Fusarium*, *Penicillium* and *Paecilomyces* sp. were greatly reduced.[128] Several other studies have indicated sensitivity of *Penicillium* sp. towards heavy metals.[34] However, some of the best examples of microbial metal tolerance are found in the genus *Penicillium*, which underlines the fact that metal responses may be species and strain specific. *Penicillium ochro-chloron* can grow in saturated $CuSO_4$ and is frequently isolated from industrial effluents, whereas *Penicillium lilacinum* comprised 23% of all fungi isolated from soil polluted by mine drainage.[45]

Heavy metals can also affect the composition of cyanobacterial, algal and protozoan populations.[96] Applications of Cu favoured coccoid, colonial cyanobacteria and inhibited *Anabaena flos-aquae*. For eukaryotic algae, Cu appeared to reduce numbers of centric diatoms, except *Skeletonema costatum*, and eliminated dinoflagellates. However, other flagellates increased and pennate diatoms, particularly *Amphiphora paludosa* v. *hyalina* became dominant components of the population.[34] In a metal-contaminated aquatic sediment only 63 taxa of benthic algae were determined whereas 111 taxa were detected over 1 km away from the main focus of infection.[81] Other studies have shown *Ankistrodesmus falcatus*, *Botryococcus braunii*, *Chlorella pyrenoidosa*, *Raphidium* sp. and *Schroederia setigera* to be more abundant in metal-contaminated waters.[6] Among protozoa and pigmented flagellates, *Peranema* and *Euglena* sp. are generally quite resistant to metal toxicants although, in comparison with walled microorganisms, protozoa are of high sensitivity. As with other organisms, changes in species composition, numbers and diversity occur with heavy metal exposure but there is considerable variation in recorded effects.[100] Thus, it seems that elevated concentrations of heavy metals can affect the qualitative and quantitative composition of microbial populations although, as mentioned, it may be difficult to separate heavy metal effects from those of other environmental components. For example, phylloplane microflora may be subject to the influence of other potential toxicants, e.g. SO_2, particularly in industrial locations, and these may contribute to observed effects.[34] In addition, there are well-known theoretical and practical difficulties in obtaining meaningful assessments of microbial numbers and diversity in natural habitats which should not be ignored.[24] Nevertheless, it is generally accepted that heavy metal induced changes in species composition do occur and a commonly associated assumption is that such changes lead to the establishment of a 'resistant' or 'tolerant' population.

Heavy metal tolerance

The ability of a microorganism to survive and reproduce in a metal-contaminated habitat may depend on genetical adaptation and/or physiological adaptation.[4,32,73,74] In a given population, both may occur to varying extents, e.g. when microorganisms are 'trained' to tolerate increasing concentrations of heavy metals in laboratory media by repeated subculture on metal-containing media, there may be physiological changes which may be unstable, and also the selection of genetically stable variants from the original population.[4,126] Obviously, 'resistance', 'tolerance' and 'sensitivity' are arbitrary terms and such distinctions are often based on fairly subjective criteria such as the ability to grow on a certain metal concentration in media. Both 'tolerance' and 'resistance' are used widely without clear distinction although it may be more appropriate to use 'resistance' to describe a direct response resulting from metal exposure. This is not an easy distinction to make and not always has there been adequate consideration of toxicity modification by environmental factors or media components which may lead to the isolation and identification of apparently resistant isolates.[47] Nevertheless, selection and isolation of metal-resistant bacteria, algae and fungi (some mentioned previously) has been described from heavy metal polluted soils and waters.[6,36,96]

As mentioned, designation of 'tolerance' or 'resistance' is arbitrary and whether or not an isolate grows in or on metal-containing media does not necessarily reveal whether any adaptation has taken place in the environment. The ability to grow in the presence of relatively high metal concentrations is found in a wide range of microbial groups and species, including those from unpolluted sites, and not in all cases is any adaptation necessary. Tolerance may result from intrinsic properties of the organism, e.g. the possession of extracellular mucilage or polysaccharide or an impermeable cell wall.[47] In addition, environmental components may have considerable influence on toxicity and therefore apparent resistance. A good example is provided by those fungi that can grow in saturated $CuSO_4$ ($\simeq 1.3$ M), and very high concentrations of other heavy metals. Such solutions are very acidic and these organisms are sensitive to sub-millimolar levels at pH values around neutrality.[48,52] A range of other examples are available which illustrate the difficulties involved. Although fungal abundance and species diversity were apparently reduced in Zn-polluted soil, there was little difference in zinc tolerance between fungi isolated from control or polluted sites and most achieved 50%

growth at 700 μM Zn^{2+}. At control sites, *Bdellospora*, *Verticillium* and *Paecilomyces* sp. were Zn-tolerant with *Aureobasidium* and *Penicillium* sp. at polluted sites.[69] Similar findings were obtained from a Cu, Ni, Fe and Co-polluted soil where fungal populations were not significantly different at polluted or control sites and tolerance (defined as growth on about 1.6 mM Cu and/or Ni in media) was displayed by fungi from both sites with *Penicillium* sp. comprising 60% of tolerant isolates followed by *Trichoderma*, *Rhodotorula*, *Oidiodendron*, *Mortierella* and *Mucor* species.[42] Although *A. pullulans* can become a dominant organism on metal-polluted phylloplanes, adaptive changes were not necessarily involved and the ability to tolerate high levels of Pb occurred in isolates from unpolluted as well as polluted sites.[82] Hallas and Cooney[57] concluded there was no significant correlation between the tin concentration in sediments and the proportion of organisms in a community resistant to tin in laboratory media: tin alone did not select for resistance.

Thus, considerable theoretical and practical difficulties may be encountered in ecological studies of metal pollution. Laboratory studies under defined conditions allow more detailed investigation of the microbial mechanisms involved in metal resistance at the cellular level and these will now be considered. However, it should be stressed that, although in some circumstances field and laboratory observations can be complementary, extrapolation of laboratory data to the field situation should be done with caution.[24]

EXTRACELLULAR COMPLEXATION

Many bacteria produce large amounts of extracellular polysaccharides that have anionic properties and thus function as efficient biosorbents for metal cations.[75,109] Such extracellular polymers, e.g. those produced by *Zoogloea* sp., are strongly involved in metal removal from sewage treatment processes and the extraction and removal of polymers from these and other bacterial cultures can greatly reduce biosorption capacities and also increase metal sensitivity.[17] The extracellular matrices act as an efficient barrier and prevent significant entry of metal ions into the cell. Similar interactions are also likely for those cyanobacteria, algae and fungi that produce extracellular polymers.

Many organic metabolites can be important in metal detoxification because of chelating or complexation properties. Citric acid is an efficient metal chelator whereas oxalic acid can precipitate metals as insoluble oxalates around cell walls and in the external medium. Oxalic acid producing fungi often exhibit marked metal tolerance and this detoxification mechanism is often found in wood-rotting fungi, particularly those exposed to copper–chrome–arsenate food preservatives, e.g. *Poria* as well as other species such as *Aspergillus niger*, *Penicillium spinulosum* and *Verticillium psalliotae*.[84] Iron is an essential element and many microorganisms release various iron-binding molecules called siderophores which complex Fe^{3+}. The externally formed ferric chelates subsequently interact with cells so that iron accumulates intracellularly.[89,97] Some of the siderophores may strongly chelate other metals, e.g. gallium, nickel, and analogous compounds were produced during growth of *Pseudomonas aeruginosa* in the presence of uranium or thorium.[1,59]

Iron limitation can increase the extracellular production of siderophores and, in *Anabaena* sp., these can function as strong copper-complexing agents.[29] It is therefore conceivable that in some circumstances a secondary manifestation of siderophore excretion may be some protection from metal toxicity.

EXTRACELLULAR PRECIPITATION AND CRYSTALLIZATION

Certain organisms, particularly sulphate-reducing bacteria, e.g. *Desulphovibrio*, are involved in the formation of sulphide deposits which contain large amounts of metals.[70] Sulphide formation thus leads to metal removal from solution and this is associated with resistance in a variety of microbes. Metal-resistant strains of *Klebsiella aerogenes* precipitate Pb, Hg or Cd as insoluble sulphide granules on outer surfaces of cells,[2] and particles of Ag_2S are deposited on thiobacilli when grown in silver-containing sulphide leaching systems.[95]

Strains of the green alga *Cyanidium caldarium* can grow in acidic waters at 45 °C containing high concentrations of metal ions. Iron, copper, nickel, aluminium and chromium can be removed from solution by precipitation at cell surfaces as metal sulphides. Cells can contain up to 20% metal on a dry weight basis.[130] Yeasts can also precipitate metals as sulphides in and around cell walls and colonies may appear dark brown in the presence of copper.[4,45]

Microbial sulphide production is also important in the biological cycle of mercury. Hydrogen sulphide is effective at volatilization and precipitation of mercury through disproportionation chemistry in the aqueous environment. Such reactions lead to mobilization of metals from the aqueous environment into the atmosphere. A key microorganism is *Desulphovibrio* and such processes occur when these bacteria are able to reduce sulphate, under anaerobic conditions, in polluted waters. The disproportionation of organomercury and organolead compounds to more volatile substances and insoluble metal sulphides by H_2S can be represented by the following reaction:

$$2CH_3Hg^+ + H_2S \longrightarrow (CH_3)_2Hg + HgS$$
$$2(CH_3)_3Pb^+ + H_2S \longrightarrow (CH_3)_4Pb + (CH_3)_2PbS$$

Once in the atmosphere, volatile organometallic compounds are unstable as metal–carbon bonds are susceptible to photolytic cleavage.[130]

Many other examples of crystallization and precipitation on microbial surfaces are known and some of these may represent resistance mechanisms. Microbes are implicated in the formation of ferromanganese nodules on ocean floors and several bacteria, e.g. *Hyphomicrobium*, algae and fungi promote Mn^{2+} oxidation in a variety of habitats and can become encrusted with manganic oxides. Other bacteria can become encrusted with oxidized iron compounds by metabolism-dependent and -independent processes.[70] In *Thiobacillus ferroxidans* and certain algae, Au^{3+} can be adsorbed to cell walls and the plasma membrane and be reduced to particles of elemental Au^0. Reducing compounds produced by silver-resistant bacteria can reduce Ag^+ to metallic Ag^0, which results in silver deposition on glass surfaces of culture vessels or in growing colonies.[11] In metal-resistant strains of *Citrobacter*, a major mechanism of metal uptake is due to the activity of a cell bound phosphatase, induced by growth in glycerol-2-phosphate, which can precipitate Cd, Pb, Cu and U as insoluble metal phosphates on the cell surface.[76,77]

The majority of phenomena just described are intimately dependent on metabolic processes and, in polluted environments, may be obvious determinants of tolerance and survival. However, such effects may be largely fortuitous and should not be considered true mechanisms of resistance, particularly when an organism excretes a potential detoxifying substance in the presence or absence of potentially toxic metal concentrations.

General biosorption of heavy metals to cell walls may obviously reduce metal concentrations in solution but the extent of this is largely outside metabolic control and dependent on such factors as metal and biomass concentration, pH, competing anions and cations, etc.[44,45,47] Biosorption may be important in certain survival structures, e.g. fungal chlamydospores which have thick cell walls, frequently melanized. Such cell types exhibit higher metal tolerance than hyaline cell types and the wall appears to act as a permeability barrier preventing entry into the cell.[44,45,53]

TRANSPORT-RELATED MECHANISMS OF RESISTANCE

Metabolism-dependent intracellular transport may be a slower process than binding and is inhibited or halted by metabolic inhibitors, low temperatures, the absence of an energy source (e.g. glucose in heterotrophs, light in phototrophs) and uncouplers.[19] In several algae, bacteria and yeasts, amounts accumulated by transport may greatly exceed amounts taken up by general biosorption.[46,70] Rates of transport are also affected by the metabolic state of the cell and the nature and composition of the growth medium. Integral to heavy metal transport systems are ionic gradients across the cell membrane, e.g. H^+ and/or K^+, and the membrane potential. Fuller treatments of transport have already been published.[19,45,49,103,106,118]

A relationship between heavy metal transport into microbial cells and toxicity is often observed, with sensitive strains taking up more metal ions than resistant strains. It is therefore not surprising that decreased transport, impermeability or the occurrence of metal efflux systems constitute resistance mechanisms in many organisms (Fig. 7.1).

In several bacteria, including *Staphylococcus aureus* and *Bacillus subtilis*, active transport of Cd^{2+} depends on the cross-membrane electrical potential and this uptake system is highly specific for Mn^{2+} as well as Cd^{2+}. Resistance may arise from reduced Cd^{2+} transport in Gram-negative bacteria.[105] In *S. aureus*, an energy-dependent Cd^{2+} efflux system, which is a $Cd^{2+}/2H^+$ antiport, is present in resistant strains which results in a net reduction in Cd^{2+} uptake[10,104,105,118] (Fig. 7.1a). Cadmium-resistant *Pseudomonas putida* also exhibits reduced uptake and this may also be a result of defective transport and/or the presence of a Cd^{2+} efflux system. In an *Alcaligenes* sp., adaptation to Cd^{2+} resistance was associated with the synthesis of a new membrane protein which may be involved in the prevention of Cd^{2+} influx or the cause of Cd^{2+} efflux. Membrane proteins that reduce bacterial uptake of Hg^{2+} have also been described. Some Cu^{2+} resistant strains of *E. coli*

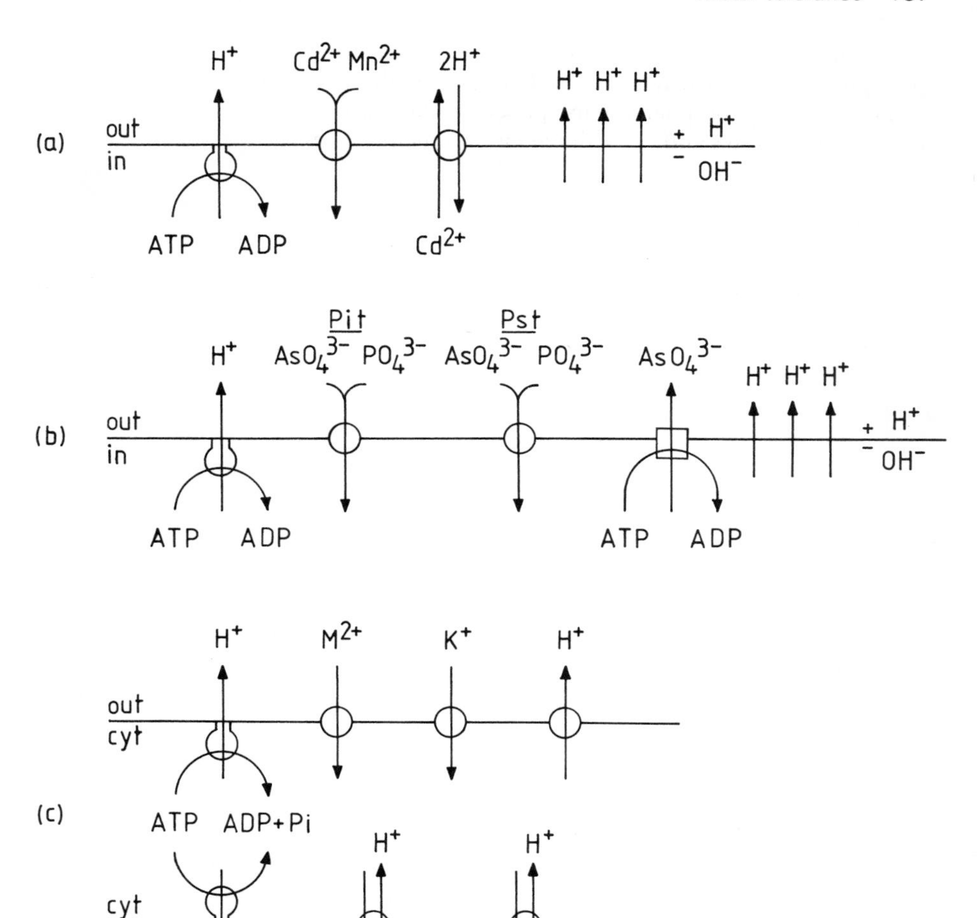

Fig. 7.1 (a) Model for Cd^{2+} uptake and efflux systems in *Staphylococcus aureus*. (b) The phosphate (arsenate) transport systems and the arsenate efflux system of *Escherichia coli*. The K_m for PO_4^{3-} is 0.25 μM for the *Pst* (phosphate specific transport) phosphate transport system and 25 μM for the *Pit* (P_i transport) phosphate transport system. The K_i for AsO_4^{3-} as a competitive inhibitor of both systems is 25 μM. (c) Suggested scheme of divalent cation (M^{2+}) and K^+ transport, driven by the H^+ gradient, into the cytoplasm and vacuole of *Saccharomyces cerevisiae*. The K^+ gradient may also be important, see ref. 93. Models adapted from refs 44, 93, 104–06, 118.

exhibit reduced uptake and/or exclude Cu^{2+} due to the disappearance of outer membrane proteins which may be involved in Cu^{2+} transport.[10]

The mechanism of arsenate resistance in *S. aureus* and *E. coli* is reduced accumulation by resistant strains which is a result of an energy-dependent efflux mechanism (Fig. 7.1b). Arsenate functions as a phosphate analogue and is generally accumulated by phosphate transport systems. Thus, as with Cd^{2+}, which can enter certain bacteria via a Mn^{2+} system, toxic anions and cations may be accumulated by means of existing transport systems for essential nutrients. However, energy-dependent efflux mechanisms, such as those for Cd^{2+} and AsO_4^{3-}, must be highly specific for the toxic anion or cation to prevent detrimental loss of the essential nutrient.[104,105] Other mechanisms of resistance to arsenicals include enzymic transformations (see later).

In *S. cerevisiae*, a connection between toxicity and energy-dependent intracellular uptake has been shown for a variety of metals including Cd, Cu, Co and Zn with decreased influx occurring in tolerant strains.[45,126] Although such reduced uptake is a feature encountered in many strains of bacteria, fungi and algae, there are several examples where resistant strains take up more metal than sensitive strains, e.g. Cu^{2+}- and Ni^{2+}-tolerant strains of *Scenedesmus* sp. and Mn^{2+}-resistant strains of *S. cerevisiae*. This may be due to the presence of additional mechanisms such as more efficient internal detoxification in the resistant strains.[45]

Impermeability has already been mentioned as a result of cell wall complexation, extracellular precipitation, etc. and these may be of benefit in reducing or preventing cellular entry of toxic metal ions. In the Cu^{2+}-tolerant fungus *Penicillium ochro-chloron*, growth in high Cu^{2+} concentrations induced the synthesis of high concentrations of intracellular glycerol which probably assisted in cellular exclusion of Cu^{2+}.[50]

Since a decreased influx can often determine tolerance, it follows that abiotic factors which reduce transport may also enable microbial growth and survival, e.g. ion competition and pH.[47] Ca^{2+} can reduce Cd^{2+} toxicity to yeast by depressing rates of influx.[71] Although the fungi *Scytalidium* and *P. ochro-chloron* can grow in saturated $CuSO_4$ at pH 0.3–2.0, they may be sensitive to $\leqslant 4 \times 10^{-5}$ M near neutrality because of enhanced Cu^{2+} uptake at such pH values.[45,52]

INTRACELLULAR COMPARTMENTATION AND DETOXIFICATION

Once inside cells, metal ions may be compartmentalized and/or converted to more innocuous forms. Such processes can be effective detoxification mechanisms and microbes expressing them may be able to accumulate metals to high intracellular concentrations. It has been suggested that such mechanisms may be temporary and precede other means of expulsion of accumulated metals from the cells, possibly by means of vacuoles in eukaryotic microbes, although there is little detailed information in this area.[130]

A variety of electron dense deposits, many of unknown composition, have been recorded in cyanobacteria, bacteria, algae, protozoa and fungi after exposure to heavy metals and radionuclides.[18,47,130] Polyphosphate has been implicated as a metal sequestering agent in *Pseudomonas putida*,[62] *Anabaena cylindrica* and *Plectonema boryanum*,[66] a variety of eukaryotic algae[123] and certain fungi and yeasts.[44,45]

In eukaryotic microbes, e.g. algae, yeasts and fungi, there may be compartmentation of accumulated metals within specific organelles. In yeasts and fungi, a major proportion of accumulated Co^{2+}, Mn^{2+}, Mg^{2+}, Zn^{2+} and K^{+} is located in the vacuole where it may be in ionic form or bound to low molecular weight polyphosphates[93,126,127] (Fig. 7.1c).

A common metal-induced response in many microorganisms is the synthesis of intracellular metal-binding proteins which function in detoxification and also the storage and regulation of intracellular metal ion concentrations. Metallothioneins are the most widely studied and these are low molecular weight, cysteine-rich proteins that can bind metals like Cd, Zn and/or Cu. A Cd-binding prokaryotic metallothionein (molecular weight approx. 8100) was first described in the cyanobacterium *Synechococcus* sp.[94] and others have since been reported in several bacteria including *P. putida*.[62] Inducible Cd-binding proteins also occur in *E. coli* but these are larger than metallothioneins and may be more related to recovery from cadmium toxicity.[80]

In fungi, the copper-induced metallothionein of *S. cerevisiae* (molecular weight approx. 6573) has received most attention[26] although a copper-inducible metallothionein has also been described from *Neurospora crassa*.[83] The yeast protein is inducible only by copper and not cadmium or zinc and is therefore often referred to as Cu-MT or yeast MT.[26] Inducible Cd-binding proteins have been isolated from *Schizosaccharomyces pombe* that are structurally different to Cu-MT. A proposed name for these sulphur-rich metal-binding polypeptides is *phytochelatin*. The metal-binding peptides are composed of only three amino acids, L-cysteine, L-glutamic acid and glycine and are analogous to similar peptides found in plant cells exposed to heavy metals like Cd, Cu, Hg, Pb and Zn.[26] The general structure of two peptides found in Cd-exposed *S. pombe* are $(\gamma\text{-Glu-Cys})_n\text{Gly}$ ($n=2$ and 3).[56] The unit peptides are called cadystin and the phytochelatins $n=2$ and $n=3$ are called cadystin A and B, respectively. In addition to these, five homologous peptides with chain lengths from $n=4$ to $n=8$ are also observed in *S. pombe*, indicating that these are synthesized by elongation of the peptide with one (γ-Glu-Cys) unit at the expense of and possibly starting from glutathione.[26,56] These cysteine-rich peptides of *S. pombe* can sequester several transition metals and therefore share some basic features with metallothioneins. Multiple enzymes may be involved in phytochelatin synthesis and therefore mechanisms must exist for metal-dependent induction or activation of such enzymes.[26]

Low molecular weight cadmium-binding proteins have been detected in Cd-resistant strains of *S. cerevisiae*. In a sensitive strain, most internal Cd was bound to insoluble cytosolic material whereas Cd-binding proteins (molecular weight $<30\,000$) were detected in the cytosol of resistant strains.[68] Other metal binding proteins of unknown structure have been described from several other filamentous fungi and yeasts. Copper-binding proteins (molecular weight approx. 8000), similar to mammalian and yeast MT, have also been found in copper-tolerant *Scenedesmus*, and metallothionein-like proteins were induced by cadmium in *Chlorella pyrenoidosa* and *Dunaliella*.[46,61]

A metallothionein-like protein has been identified in the ciliate *Tetrahymena pyriformis* after exposure to cadmium, which is also capable of zinc binding. The protein is rich in cysteinyl residues (32.4%) and acidic amino acids (23.7%) and contained 3.7, 0.7 and 0.1 g atom Cd, Zn or Cu, respectively.[86] Two kinds of Cd-binding proteins were detected in *Amoeba proteus*, with molecular weights of >45 000 and 12 000, the latter being analogous to mammalian metallothionein.[3]

METAL TRANSFORMATIONS

Microorganisms can carry out chemical transformations of heavy metals such as oxidation, reduction, methylation and demethylation and these are relevant not only to biogeochemical cycling but may also constitute mechanisms of resistance. Most work on metal transformations has concentrated on bacterial involvement in the mercury cycle which can be simply represented as:

$$CH_3Hg^+ \underset{\text{methylation}}{\overset{\text{demethylation}}{\rightleftharpoons}} Hg^{2+} \underset{\text{oxidation}}{\overset{\text{reduction}}{\rightleftharpoons}} Hg^0$$

Mercury resistance is a common property of Gram-positive and Gram-negative bacteria, the determinants usually being plasmid encoded particularly in Gram-negatives (see later).[39,40] The most ubiquitous mechanism of mercury resistance in bacteria is the enzymatic reduction of Hg^{2+} by cytoplasmic mercuric reductase to metallic Hg^0 which is less toxic than Hg^{2+}, volatile and rapidly lost from the environment[40,65,105,111,129] (Fig. 7.2). This has also been recorded in certain fungi and yeasts.[45,47] Organomercurial compounds are enzymatically detoxified by organomercurial lyase which cleaves the Hg–C bond of, e.g. methyl-, ethyl- and phenyl mercury, to form Hg^{2+} and methane, ethane and benzene, respectively. The Hg^{2+} can then be volatilized by mercuric reductase. Thus

Organomercurial lyase

$$C_6H_5Hg^+ \longrightarrow C_6H_6\uparrow + Hg^{2+}$$
Phenyl mercury → benzene

$$CH_3Hg^+ \longrightarrow CH_4\uparrow + Hg^{2+}$$
methyl mercury → methane

$$C_2H_5Hg^+ \longrightarrow C_2H_6\uparrow + Hg^{2+}$$
ethyl mercury → ethane

Mercuric reductase

$$Hg^{2+} \xrightarrow{\text{NADPH} \rightarrow \text{NADP}} Hg^0\uparrow$$

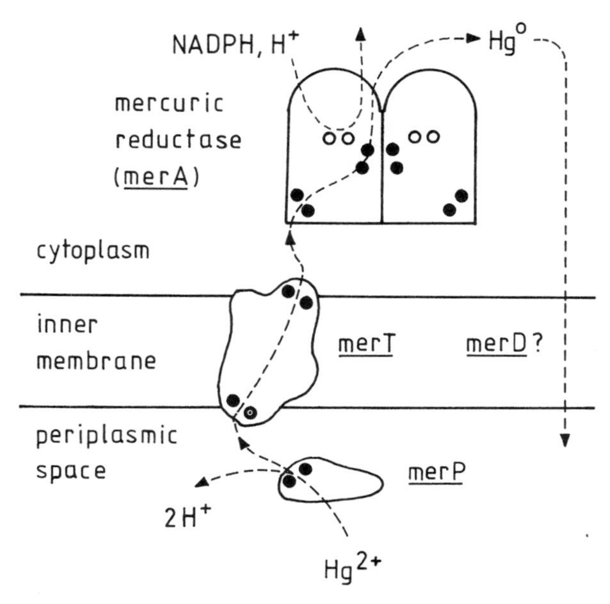

Fig. 7.2 The model for bacterial mercury resistance. The suggested locations of the *merP*, *merT* and *merA* gene products are shown relative to the inner membrane of the bacterial cell. Solid circles represent the paired cysteine residues which bind the Hg^{2+} ion. The mercuric reductase protein is represented as a dimer. The open circles represent the redox-active cysteines (and the FAD) at the active site of the reductase. Adapted from refs 25 and 106.

More than one kind of organomercurial lyase may be present in a given bacterium. In organomercurial resistant *Escherichia coli*, two enzymes were active against phenylmercuric acetate whereas only one was active against methyl- and ethylmercury chloride. In a *Pseudomonas* strain there were also two enzymes of similar specificities as well as being able to cleave *p*-hydroxymercuribenzoate.[105,129]

FAD-containing mercuric reductase is a flavoprotein and is the best studied metal detoxification enzyme and this, and organomercurial lyase, requires an excess of thiols for activity. NADPH is the preferred reducing co-factor for mercuric reductase, as well as organomercurial lyase, but in some bacterial strains NADH is effective.[129] The thiols prevent the formation of NADPH–Hg^{2+} complexes and ensure that the Hg^{2+} is present as a dimercaptide. The reaction catalysed by mercuric reductase *in vitro* can be represented:

$$RS{-}Hg{-}SR + NADPH + H^{+} \longrightarrow Hg^{0} + NADP + 2RSH$$

Mercuric reductase is structurally and mechanistically related to glutathione reductase and lipoamide dehydrogenase.[40,105] The mechanism of mercuric reductase is probably that electrons are transferred from NADPH via FAD to reduce the active site cystine, converting it to two cysteine residues with titratable

SH groups. One Cys residue forms a charge transfer complex with FAD. The active site cysteines then reduce Hg^{2+} bound to the C-terminal cysteines, forming Hg^0.[40,105,129] Mercuric reductase has been reported to exist as a monomer, dimer and a trimer with a sub-unit molecular weight of between 54 000 and 69 000 depending on the source,[129] though it seems probable that the dimeric structure is active *in vivo*[40] (Fig. 7.2).

In contrast to the processes of demethylation of CH_3Hg^+ to Hg^{2+} and the reduction of Hg^{2+} to Hg^0, much less is known about the reverse processes, the oxidation of Hg^0 to Hg^{2+} and the methylation of Hg^{2+} to CH_3Hg^+ and $(CH_3)_2Hg$, although microbes are implicated in both reactions.[105,114,129]

A wide range of common bacteria including *E. coli*, *Pseudomonas fluorescens*, *Bacillus subtilis* and *B. megaterium* have been reported to oxidize Hg^0 to Hg^{2+}, although such bacteria do not generally carry out subsequent methylation. It has been suggested that catalase is responsible for such Hg^0 oxidation.[105,129] Since this reaction results in formation of more toxic Hg^{2+}, it cannot be regarded as a resistance mechanism. In addition, the organic content of the medium greatly influences apparent Hg^0 oxidation and, since the direct interaction of bacterial cells was not essential, microbial Hg^0 oxidation needs to be viewed with caution.[65]

Methylation of mercury can be accomplished by abiotic and biotic means; microbes capable of methylation include bacteria, fungi and yeasts from a range of habitats, e.g. soil, water, sediments.[114] Although a more toxic mercury compound results, methylation may be viewed as a detoxification mechanism since methylated species are more volatile and may be lost from the environment.[30,47,114,130] In addition, the cellular elimination of organometallic compounds may be easier by diffusion.[130] There is general agreement that methylation of Hg^{2+}, by direct and indirect microbial action, involves methylcobalamin (vitamin B12).[105,130] There appears to be two mechanisms for methyl transfer from methylcobalamin to mercury and other heavy metals; first, electrophilic attack by the metal on the Co–C bond of methylcobalamin, e.g. Hg^{2+}, Pb^{4+}, Tl^{3+} and Pd^{2+}, and, second, methyl-radical transfer to an ion pair between the attacking metal ion and the portion of the methylcobalamin molecule that complexes to the central Co atom, e.g. Pt^{2+}/Pt^{4+}, Sn^{2+}, Cr^{2+}, Au^{3+}.[130] The ecological significance of methylcobalamin-dependent methylation as a detoxification mechanism is illustrated by two strains of *Clostridium cochlearium*, one possessing and the other lacking such a system. Both strains transport Hg^{2+} into cells at the same rate but the methylcobalamin-independent strain is much more Hg^{2+} sensitive.[130]

Other detoxifying enzymes include arsenite oxidase which catalyses $As^{3+} \longrightarrow As^{5+}$ and chromate reductase which catalyses $Cr^{6+} \longrightarrow Cr^{3+}$. The toxicity of both these elements depends on their chemical form and oxidation state. Thus, trivalent As^{3+} is 100–200 times more toxic than As^{5+}, and so bacterial oxidation of arsenite to arsenate may be an effective detoxification mechanism. The most studied enzyme is that of a strain of *Alcaligenes faecalis* which has a high affinity for arsenite ($K_m = 2.2$ μM). Chromates (CrO_4^{2-}) and dichromates ($Cr_2O_7^{2-}$) which contain chromium with a valency of +6 are more toxic than those compounds which contain the reduced form of chromium, Cr^{3+}. Enzymatic reduction of Cr^{6+} by chromate reductase has been reported in a chromate-resistant strain of

Pseudomonas fluorescens.[129] Oxyanions of selenium and tellurium are generally of high microbial toxicity and the reduction of these may result in detoxification. Corresponding metallic compounds have been found associated with resistant bacterial cells when grown with tellurite or selenite, although the enzymes responsible have not been characterized.[129] It is likely that many other examples of microbially catalysed metal transformations will be discovered with further research.

MORPHOLOGICAL CHANGES

A variety of heavy metal induced morphological changes have been recorded in microorganisms but often it is not clear whether these represent survival mechanisms or are manifestations of toxicity.[45,108] In some cases, morphological effects may be connected to possible roles of essential metal ions in morphogenesis. The yeast–mycelium (Y–M) transition of *Candida albicans* can be suppressed by Zn^{2+} concentrations $< 10\ \mu M$ and this effect is temperature dependent, itself an important determinant of morphological transitions of *C. albicans*. Micromolar Zn^{2+} concentrations can inhibit mycelial development at 25 °C but not at 37 °C, even at millimolar concentrations.[8]

Spore-bearing hyphal aggregates called synnema (= coremia) are concerned with the spread and survival of certain fungi and a variety of external stimuli trigger their development, including metal compounds. During growth of *Penicillium funiculosum* in the presence of potentially toxic tributyltins, synemma production resulted. Aggregation of hyphae resulted in formation of primordia which rose above the colony and, after thickening and apical growth, bore large numbers of conidia.[87] Such a development results in a wider spatial separation of the conidia and substrate and not only aids dispersal but also ensures conidium formation away from potential toxicants in the substrate. If synnema formation is a survival strategy, it may be of wider significance in fungi exposed to inorganic and organometallic compounds.[87]

In *Aureobasidium pullulans* and certain other fungi, e.g. *Phoma glomerata*, inorganic and organic metal compounds induce or accelerate the formation of melanin-pigmented hyphae and chlamydospores. Melanized cell structures are impermeable to heavy metals and more metal tolerant than hyaline cell types.[44] Since a major proportion of fungal biomass is present as melanized forms in terrestrial environments,[9] this may be a tolerance mechanism of ecological importance.

Commercial aspects

There are commercial aspects to several areas of metal–microbe interactions. Heavy metal toxicity is the basis of many antimicrobial preparations for the control of animal and plant pathogens, algal blooms, and the preservation of natural and synthetic materials.[41,43] Further knowledge of the mechanisms of toxicity and resistance may be useful for the formulation and application of more effective compounds.[43]

As previously described, many interactions of heavy metals with microbes result in their removal from solution. These processes are of commercial relevance because the removal of potentially toxic heavy metals and radionuclides from industrial effluents and waste waters can lead to detoxification and/or the recovery of valuable metals, e.g. gold.[22,44,45,101,121] The use of symbiotic microbial systems, e.g. mycorrhizas, for polluted land reclamation may also be a commercial possibility.[20]

The industrial potential of microbial metal recovery systems depends on many factors, including capacities, efficiencies and selectivity, the ease of desorption, equivalence to existing physical and chemical processes in performance and economics and immunity from interference, by other effluent components or operating conditions. For full competition with existing technologies, it has been suggested that removal efficiencies must be >99% and loading capacities >150 mg metal g^{-1}.[22,23] An advantage of microbial systems for commercial use is their versatility in respect of uptake mechanisms and the relative ease of morphological, physiological and genetical manipulation. Both living and dead microbes can take up heavy metals and radionuclides as can products produced by or derived from microbial biomass, e.g. excreted metabolites, polysaccharides and cell wall constituents.[22,23,46,70] There is no doubt that certain kinds of biomass or products can be highly efficient accumulators, even from dilute external concentrations, and uptake capacities can be large by means of a variety of mechanisms ranging from purely physico-chemical to those dependent on cell metabolism (Table 7.2). Microbe-based technologies may provide an alternative or supplementary process to conventional treatment methods of metal removal/recovery and several are in commercial operation in the mining and metallurgical industries.[46,64] However, it is only recently that biosorptive systems are receiving greater sophistication and many areas of metal/microbe interactions remain unexplored.

As described previously, heavy metals and related elements can be toxic and this may hamper industrial applications of living cells. However, it may be possible to separate cell propagation from the metal contacting phase or to use strains tolerant to the metal concentrations encountered. Clearly, resistance is a widely found property in microbes. The use of dead biomass or derived products eliminates problems of toxicity, nutrient supply and maintenance of optimal growth conditions. However, living cells can exhibit a wider variety of mechanisms for metal accumulation including transport, intracellular and extracellular precipitation, etc. although, as previously discussed, resistance may be associated with decreased uptake.

The uptake of heavy metals by microbes can often be separated into a phase of metabolism-independent binding or adsorption to cell walls and other external surfaces followed by metabolism-dependent transport into cells. In some instances, intracellular uptake may be a result of increased cell permeability, particularly where toxicity is manifest. Metabolism-independent processes of metal uptake, including adsorption, can all be termed 'biosorption', are frequently rapid and can occur in dead as well as living cells. In many organisms, binding may be at least a two-stage process, first involving interactions between metal ions and reactive groups followed by inorganic deposition of increased amounts of metal which leads to much greater amounts of metal being accumulated than by purely ion-

Table 7.2 Metal accumulation by microorganisms[a]

Organism	*Element*	*Uptake (% dry weight)*
Bacteria (170 strains)	Cd	0–2
Bacteria (19 strains)	Ag	0.7–4.4
Bacteria (3 strains)	U	8–9
Actinomycetes (5 strains)	U	8–9
Streptomyces sp. (12 strains)	U	2–14
Citrobacter sp.	Pb	34–40
	Cd	13.5
Pseudomonas aeruginosa	U	15
Zoogloea sp.	Cu	34
	Cd	40
	Co	25
Mixed bacterial culture	Ag	32
Mixed bacterial culture	Cu	30
Chlorella regularis[b]	U	15
Chlorella vulgaris[b]	Au	10
Scenedesmus obliquus	Cd	0.3
Euglena sp.	Al	0.2–1.8
Fungi (47 strains)	Cu	<0.05–0.3
Fungi (20 strains)	Ag	0.7
Phoma sp.	Ag	2
Rhizopus arrhizus	Cu	1.7
	Pb	10.4
	Ag	5.4
	U	19.5
	Th	18.5
Penicillium sp.	U	8–17
Yeasts (14 strains)	Cu	<0.05–0.2
Yeasts (6 strains)	Ag	0.05
Saccharomyces cerevisiae	Cd	0.2–3.1
	Zn	0.5
	U	10–15
	Th	11.6

[a] Data obtained from a number of sources, particularly refs 11, 28, 31, 33, 46, 54, 76–9, 85, 90, 101, 110, 118, 119. Referral to original references is recommended for experimental details such as initial metal concentrations, pH, incubation times and cell densities. In this table, uptake refers to total uptake and may include processes dependent and independent of metabolism
[b] Immobilized algal cells

exchange phenomena.[12,13,70] For some metals and radionuclides, e.g. Pb, U, Th, and in certain organisms, particularly filamentous fungi, most accumulation is composed of surface phenomena.[44,46,54] It should be stressed that in growing cultures, biosorption and transport can be obscured or enhanced by other phenomena, e.g. precipitation, complexation, changes in the growth medium and morphology of the organism.[45,47]

Many potential binding sites occur in microbial cells and extracellular matrices, and structural components in bacteria, algae and fungi include peptidoglycan, teichoic and teichuronic acid, polysaccharides, cellulose, uronic acid, proteins, glucans, mannans, chitin and melanin. A variety of ligands may be involved in biosorption including carboxyl, amino, phosphate, hydroxyl, carbonyl and sulphydryl groups and the relative importance of each is difficult to resolve.[110] Since the chemical composition of microbial cell walls can vary considerably, there are considerable differences in adsorption capacities between species, strains and even different cell types of the same organism.[44] In general, bound metal, unless crystallized, is relatively easy to remove and binding is not generally affected over modest ranges of temperature.[44,45] However, such factors as the organic and inorganic composition of the reaction mixture, pH and the biomass density are important determinants of biosorptive processes.[44–47]

An important consideration in microbe-based methods of metal accumulation is the ease of metal/radionuclide recovery either for reclamation or for further containment of toxic/radioactive waste. Non-destructive recovery may be desired for regeneration of the biomass for re-use in multiple cycles.[119] Destructive recovery may be by pyrometallurgical treatments or dissolution in strong acids or alkalis.[22] If cheap, waste biomass is employed for valuable metals; destructive recovery may be economically feasible. As well as economic considerations, the choice of a recovery process also depends on the mechanism of accumulation. Metabolism-independent biosorption is frequently reversible and lends itself to non-destructive desorption. However, metabolism-dependent intracellular accumulation and related processes are largely irreversible and necessitate destructive recovery.[46] Not surprisingly, most attention has focused on non-destructive desorption. In general, bicarbonates have possibly the most commercial potential for recovery of radionuclides like uranium.[119] Other desorption agents include dilute mineral acids, carbonates and organic chelating agents and, in some instances, selective elution schemes have been identified.[31]

As well as dissolved metal forms, particulate material can also be adsorbed onto microbial cells. Copper, lead and zinc sulphides were taken up by mycelium of *A. niger* whereas *Mucor flavus* could take up lead sulphide, zinc dust and ferric hydroxide ('ochre') from acid mine drainage.[124]

Systems using living organisms have received most use in connection with effluents containing metals at concentrations below toxic levels. Artificial stream meanders and tailing ponds, often in series, containing photosynthetic algae, e.g. *Cladophora*, *Spirogyra*, *Rhizoclonium* and cyanobacteria, e.g. *Oscillatoria*, have been used to treat effluents from lead and uranium mining and were effective in reducing levels of Pb, Zn, Cu, Mn, U, Se, Ra and Mo to those required for permissible discharge.[55] With increasing metal concentrations and times of exposure, toxicity may result in death and settling of the biomass. Subsequent decomposition can result in H_2S production by sulphate-reducing bacteria, e.g. *Desulphovibrio*, *Desulphotomaculum*, and precipitation of the metals as sulphides. This process is virtually irreversible providing reducing conditions are maintained. Some proposed arrangements utilizing this phenomenon include single- and double-stage systems where continuous cultures of sulphate-reducing bacteria receive metal-containing inflow or H_2S from such cultures is fed into a precipitation vat that receives the metal inflow.[21,22]

Continuous mixed cultures of metal-resistant bacteria have been used for metal removal. A silver-tolerant community could tolerate up to 100 mM Ag^+ and amounts accumulated reached over 30% of the biomass dry weight.[28] Another stable, ten-membered bacterial community was isolated from activated sludge using chemostat enrichment and this could tolerate up to 15 mM Cu^{2+} and again accumulate Cu^{2+} up to 30% of the cell dry weight.[33] Activated sludge biomass is of considerable applied potential, particularly because of the role of extracellular polymers in metal accumulation. Important species include *Zoogloea* and *Arthrobacter*.[75,90]

An example where a living cell system acts as a 'polisher' to an existing process, that is not completely efficient, is the biological removal of arsenic from copper-refining effluents. After initial pyrometallurgical treatment, a two-stage algal treatment was devised with arsenic removal being accomplished in two bioreactors. Arsenic-loaded algae were subsequently harvested and incinerated.[107]

As described previously, metal transformations by resistant microbes may result in removal from solution. Continuous cultures of sewage-maintained Hg^{2+}-resistant bacteria, which catalyse $Hg^{2+} \longrightarrow Hg^0$, could remove mercury from sewage at a rate of 2.5 $mg\,l^{-1}\,h^{-1}$ (98% removal). The volatilized mercury may be collected by condensation.[58]

IMMOBILIZED CELLS

For rigorous industrial applications, freely suspended microbial biomass has a number of disadvantages. These include small particle size, low density and low mechanical strength which can limit the choice of reactor systems and make biomass/effluent separation difficult and/or expensive.[120] For use in packed bed or fluidized bed bioreactors, immobilized or pelleted biomass, living or dead, may have greater potential. Advantages include easier separation of cells from the reaction mixture, better capability for biomass re-use, high biomass loadings and high flow rates can be achieved, with or without recirculation, and there is minimal clogging in continuous flow systems. Polyacrylamide-immobilized cells of *Streptomyces albus* could remove uranium, copper and cobalt, in the order $UO_2^{2+} \gg Cu^{2+} > Co^{2+}$. Desorption could be achieved by elution with 0.1 M Na_2CO_3 and uptake capacities were not significantly affected by up to five uptake–desorption cycles.[85] A metal-resistant *Citrobacter*, immobilized in polyacrylamide, removed uranium, cadmium, copper and lead from solutions supplemented with glycerol 2-phosphate. Phosphatase-mediated cleavage of this substrate released HPO_4^{2-} which precipitated the metals around the cells as insoluble metal phosphates, e.g. $CdHPO_4$.[76–79] Polyacrylamide-entrapped *Chlorella* can successfully remove UO_2^{2+}, Au^{3+}, Cu^{2+}, Hg^{2+}, Ag^+ and Zn^{2+} from solution and a selective elution scheme for subsequent recovery is possible. Cu^{2+} and Zn^{2+} were desorbed by lowering the pH to 2 whereas Au^{3+}, Ag^+ and Hg^{2+} were not. The addition of a strong ligand at differing pH values was necessary to elute these; Au^{3+} and Hg^{2+} were selectively eluted using mercaptoethanol.[31] Other immobilization matrices include alginates and various polymeric membranes of unrevealed composition. The latter can be made to any size, contain only 10% inert material and can be used in multiple adsorption–desorption cycles.[120] Commercial processes exist for the application of such systems in fixed bed or fluidized bed reactors, particularly for recovery of valuable metals. Metal removal from dilute solutions (10–100 $mg\,l^{-1}$) exceeded 99% and some uptake values recorded, in $mmol\,g^{-1}$, were: Ag, 0.8; Cd, 1.9; Cu, 2.4; Pb, 2.9; Au, 2.0; and Zn, 2.1.[23,64] A method of using immobilized cells without entrapment is to use biofilms on inert matrices such as solid supports, foams or minerals.[72,77,79,113] Filamentous fungi can be grown in pellet form and therefore have some advantages in common with immobilized particles. Pellets (4 mm diameter) of *Aspergillus niger* were used for uranium removal in a fluidized bed reactor and were found to be more efficient than a commercially obtained ion-exchange resin. However, fungal pellets were prone to disintegration and similarities in density with the liquid medium made continuous operation difficult.[131]

Both living and dead biomass can be used in immobilized systems. If living cells are used, either entrapped or as biofilms, then it may be possible to remove other undesirable pollutants as well as heavy metals and radionuclides. Pure and mixed bacterial cultures, mainly *Pseudomonas*, have been used for simultaneous denitrification and removal of uranium and other heavy metals, the bacteria being grown as a film on anthracite coal particles in a fluidized bed or immobilized on polyvinyl chloride (PVC) or polypropylene webs.[63] Excess cells were used as a uranium biosorbent in a separate stirred tank reactor or, alternatively, were circulated contrary to uranium flow. A uranium reduction from 25 to 0.5 $g\,m^{-3}$ resulted with a mean liquid residence time of only 8 min.[102] A large-scale commercial process which treats effluents from gold mining and milling (5.5×10^6 gallons day^{-1}) uses rotating disc, biological-contacting units to simultaneously degrade cyanide, thiocyanate and ammonia. Heavy metals are taken up by the microbial film on the disc surfaces which is periodically removed for controlled disposal.[64]

A current consensus is that, for commercial use, immobilized or pelleted preparations should be used with recovery involving a cheap desorptive agent. As described, several processes are in commercial operation which are competitive in cost and operational characteristics with standard treatments. These systems may also be used in conjunction with methods relating to the removal of other impurities.[23] However, such systems have mainly been applied to valuable metals or radionuclides. Some of the microbe-based processes for general detoxification are relatively unsophisticated and use living cells, e.g. in ponds or streams. With increasing interest in microbe-based technologies for metal removal/recovery, further development of methodologies is inevitable and also further exploitation of so-far neglected interactions, e.g. particulate metal accumulation, extracellular precipitation and complexation and metallothioneins.

Genetic aspects

With a few exceptions, genetic studies of metal–microbe interactions have been limited and often confined to studies of mutant strains, mainly because knowledge

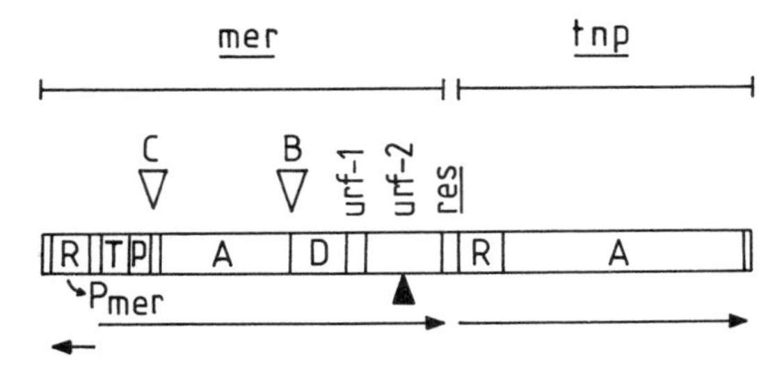

Fig. 7.3 Location of the *mer* and *tnp* genes in the mercury resistance transposon Tn*501*. The open triangles mark the positions of additional genes in other *mer* operons: C, *merC* in Tn*21* (R100); B, *merB* in the broad-spectrum operon of pDU1358 (this does not imply that the broad-spectrum operon is part of a transposon). The solid triangle is the position of the 11.2 kb insertion in Tn*21* which carries the streptomycin and sulphonamide resistance genes. P_{mer} is the promoter regulated by the *merR* gene product. The remaining genes are involved in transposition and are *res*, the internal resolution site; *tnpR*, transposition resolvase; *tnpA*, the transposase. The lower lines indicate the origin and direction of transcripts. Adapted from refs 25 and 40.

of the physiology and biochemistry of resistance is fragmentary in many cases. However, certain areas of metal–microbe interactions have received precise genetical attention and this work has not only contributed to our knowledge of microbial responses to heavy metals but also to the wider contexts of molecular biology and also biotechnology. The two areas for which considerable genetical advances have been made to date are in heavy metal (mainly mercury) resistance in bacteria and the metallothionein of *S. cerevisiae*. It is these two subjects which will be outlined in this section. It should be borne in mind that, in several other areas of metal studies, not only is genetical information limited but essential basic physiological and biochemical studies are also lacking. There appears to be much scope for further endeavour.

GENETIC ASPECTS OF HEAVY METAL RESISTANCE IN BACTERIA

Mercury resistance is a common property of many Gram-positive and Gram-negative bacteria isolated from clinical sources and the environment. The resistance determinants are usually plasmid encoded, particularly in Gram-negative bacteria.[39,40,104–106] A common feature of the Hg^{2+}-resistance determinants of bacteria from the environment is the absence of linkage to drug resistance markers in contrast to clinical bacterial isolates where Hg^{2+} resistance is usually linked to genes for antibiotic resistance.[39,40] Several mercury resistance transposons have been described that could be responsible for the widespread occurrence of Hg^{2+} resistance in Gram-negative bacteria of clinical origin.[40] To date, the most recent advances have stemmed from DNA sequencing studies of the *mer* genes located on transposons Tn*501* and Tn*21* from Gram-negative bacteria and such studies have also been carried out with broad-spectrum determinants in both Gram-negative and Gram-positive bacteria[40,106] (Fig. 7.3).

As described earlier, mercury resistance can be determined by the enzymatic reduction of Hg^{2+} to Hg^0 by mercuric reductase and this is the system on which most genetical and biochemical progress has been made.[39,40,98,106,111] However, it should be mentioned that other mechanisms of mercury resistance have been described, e.g. sulphide precipitation or reduced Hg^{2+} uptake (see earlier), but these have not been the subject of intensive genetical attention. Mercuric reductase enzymes specified by plasmids in Gram-negative bacteria can be classed in two groups based on immunological studies. The prototype type I enzyme is specified by the IncP plasmid pVS1 which carries transposon Tn*501*. The prototype for the type II enzyme is specified by plasmid R100 which carries Tn*21*. Certain other plasmids may also specify these two types of enzyme. Mercuric reductases from Gram-positive bacteria, e.g. *S. aureus*, are less heat resistant and show no immunological cross reaction with enzymes from Gram-negative bacteria.[40,104] Most work has been carried out on the Tn*501*-encoded enzyme which can constitute up to 6% of the soluble protein of *Pseudomonas aeruginosa* under conditions of Hg^{2+} induction. The enzyme encoded by plasmid R100 is very similar, the two enzymes showing 86% identity of amino acid residues and a high degree of conservative substitutions throughout their length.[25]

Two mercury resistance phenotypes occur in Gram-negative bacteria. Narrow-spectrum resistance to Hg^{2+}, and one or two organomercurials, is exhibited by strains that solely express the reductase and is widespread, whereas broad-spectrum resistance to Hg^{2+} and a larger number of organomercurials is found in strains specifying both reductase and lyase and is associated with plasmids of only a few incompatibility groups.[25,40,98] In some organisms, organomercurial resistance does not involve lyase activity. All Gram-positive Hg^{2+}-resistant staphylococci express a broad-spectrum phenotype different to that of Gram-negatives.[40] Organomercurial resistance conferred by narrow-spectrum resistance determinants is probably due to permeability and transport processes.[25,40] The archetypal *mer* operons of Gram-negative bacteria from plasmids pVSI and R100 are located on the transposons Tn*501* and Tn*21*, respectively (Fig. 7.3). Many multiresistance transposons in Gram-negative bacteria are related to Tn*21* and may have evolved from an ancestral *mer* transposon by acquisition of DNA segments specifying resistance to different drugs.[40] Tn*501* and Tn*21* are members of a class of transposable elements related to Tn*3* which specifies ampicillin resistance.

Several other Hg^{2+} resistance transposons have been described and these may include multiresistance transposons Tn*1642*, Tn*1696* and Tn*2101* as well as Hg^{2+} resistance transposons isolated from environmental bacteria, e.g. Tn*1861*, Tn*3401*, Tn*3402* and Tn*3403*. There appear to be no reports of Hg^{2+} resistance transposons in Gram-positive bacteria.[40]

Initial genetic analysis of the R100 plasmid (which contains Tn*21*) showed at least three genes were involved in Hg^{2+} resistance which correlated with the biochemical processes involved. The *merR* gene is regulatory and involved in the induction of the remaining genes, the *merT* gene encodes a protein involved in Hg^{2+} transport whereas *merA* is the structural gene for mercuric reductase[25,40,129] (Fig. 7.3). Subsequent work with R100 from a *Shigella* has revealed additional genes: *merP*, apparently required for mercury transport, and *merD* and *merC* of unknown function.[25,106] This R100 system does not encode for organomercury lyase for which the gene designation *merB* is reserved.[25] The *mer* regions of plasmid

R100 and transposon Tn*501* have been sequenced and a marked degree of similarity is evident between them with approximately 83% overall homology.[25]

Based on genetical and biochemical information, including amino acid sequences, a model for Hg^{2+} resistance has been proposed (Fig. 7.2). The mechanism described has the advantage that highly toxic Hg^{2+} is sequestered in the periplasmic space and does not encounter cellular constituents until it has been detoxified. The *merP* gene product scavenges the periplasm for Hg^{2+} which binds to a cysteine pair and releases $2H^+$. The Hg^{2+} is then transferred to the outermost cysteine pair of the *merT* gene product, also with release of $2H^+$. Subsequent Hg^{2+} transfers are all suggested to occur via such a redox-exchange mechanism. The Hg^{2+} is then transferred to the innermost cysteine pair on the inner face of the *merT* gene product. Mercuric reductase may transiently associate with the *merT* gene product in the membrane. After reduction of Hg^{2+}, elemental Hg^0 passes out through the cell envelope by simple diffusion and, since it is volatile, will be lost before it can be reoxidized.[25] Gaps in this model relate to the unknown roles of several potential *mer* gene products. The *merC* gene of plasmid R100, which has no counterpart in Tn*501*, may be responsible for some of the subtle differences between the two systems in their Hg^{2+} response. Mutants of the *merD* gene exhibit a mercury-sensitive mutant phenotype only on high-copy-number plasmids. It is believed the *merD* gene product facilitates Hg^0 diffusion out of the cell and the gene product is required only at the high values of mercury flux which occur when the *mer* genes are on a high-copy-number plasmid. The reasons why a *merD* mutant is Hg^{2+}-sensitive may be that high concentrations of Hg^0 in the cell cause product inhibition of mercuric reductase which cannot then further reduce Hg^{2+}, with obvious deleterious consequences. The low mercury flux which occurs when the *mer* genes are on a low-copy-number plasmid is adequately maintained by passive diffusion of Hg^0 out of the cell.[25]

The broad-spectrum *mer* operon from the *S. aureus* plasmid pI258 has been cloned and sequenced. Two genes (*merA* and *merB*) were identified by their homology to equivalent genes in the Gram-negative systems. They are organized in the same way as in pDU1358 which contains the broad-spectrum operon of Gram-negatives.[40] It is presumed that organomercurial compounds cross the cytoplasmic membrane via the *merP merT* system in a manner similar to Hg^{2+}. A *merA merB* deletion mutant is hypersensitive to organomercurials as well as Hg^{2+} which appears to confirm this.[25]

A variety of mechanisms of cadmium resistance are found in bacteria and some have received genetical attention. Plasmid-determined cadmium resistance is found only in *Staphylococcus aureus* and two separate genes are involved, *cadA* and *cadB*, which respectively confer a large and a small increase in Cd^{2+} resistance. The effect of *cadA* masks the *cadB* gene effect; *cadA*$^+$ *cadB*$^+$ strains are of the same resistance as *cadA*$^+$ *cadB*$^-$ strains. The mechanisms encoded are Cd^{2+} efflux for *cadA* and increased binding for *cadB*, possibly due to a membrane component analogous to metallothionein.[104–106] Cd^{2+} and Zn^{2+} resistance in *S. aureus* are genetically linked, indicating that *cadA* and *cadB* also specify Zn^{2+} resistance. However, a third Cd^{2+} resistance system in *S. aureus* involves energy-dependent Cd^{2+} efflux like *cadA* but confers Cd^{2+} resistance alone.[106] In *B. subtilis*, a chromosomal mutation is the basis of a change in the Mn^{2+} transport system, which transports both Mn^{2+} and Cd^{2+}, so that Cd^{2+} is no longer accumulated.[112]

Arsenate, arsenite and antimony(III) resistances are all coded for by an inducible operon-like system on plasmids of *S. aureus* and *E. coli*, each ion inducing all three resistances. Arsenate and arsenite resistance depend on efflux mechanisms which result in reduced net accumulation.[106] There is no evidence of plasmid involvement in arsenite oxidation by bacteria.[105] Plasmid-determined resistance to chromate has been found in *Pseudomonas* strains and other bacteria, the resistance mechanism being reduced CrO_4^{2-} uptake by plasmid-bearing strains.[106] Plasmid-determined Cu^{2+} resistance has been found on an antibiotic resistance plasmid from an *E. coli* strain, isolated from pigs fed copper feed supplements, and Cu^{2+}-treated *Pseudomonas syringae*.[106,116] A possible resistance mechanism is reduced uptake by the resistant strains.[116] Several other plasmid-determined heavy metal resistances are known in bacteria, e.g. to silver, cobalt, lead and nickel, and these are likely to receive attention in the future.[104–106,115–117]

GENETIC ASPECTS OF YEAST METALLOTHIONEIN

As previously discussed, copper resistance in *S. cerevisiae* can be mediated by the induction of a 6573 Da cysteine-rich metallothionein, termed Cu-MT or yeast MT. This metal-regulated genetic system is proving a powerful tool in pure and applied microbiology and biochemistry and the molecular and genetical basis of this response has received intensive study.

S. cerevisiae contains a single Cu-MT gene which is present in the *CUP1* locus. In genetic terms, copper resistance is ascribed to the function of the *CUP1* locus which is located on chromosome VIII[37,125] (Fig. 7.4). Copper-resistant strains *S. cerevisiae* ($CUP1^r$) contain two to fourteen or more copies of the *CUP1* locus, which are tandemly repeated (Fig. 7.4) and can grow on media containing $\geqslant$ 2mM copper.[38] Copper-sensitive strains of *S. cerevisiae* ($cup1^s$) do not grow on media containing $\geqslant$0.15 mM[26] although such a differential growth response of resistant and sensitive strains will depend on the particular medium employed. In addition, continuous subculture of yeasts in the presence of increasing copper concentrations can result in selection of hyper-resistant yeast strains[126] which appear disomic for chromsome VIII which contains the *CUP1* locus. The disomic chromosomes exhibit differential *CUP1* gene amplification patterns and this indicates that the mechanisms of copper resistance involves not only amplification of the *CUP1* locus on the chromosomes but also disomy or aneuploidy of chromosome VIII.[26] The DNA fragments of the *CUP1* locus that confer copper resistance and encode for Cu-MT have been cloned.[27,37,60] Evidence that the *CUP1* locus was present as multiple copies in $CUP1^r$ strains was obtained after restriction enzyme analysis of the cloned DNA. The basic repeating unit '*CUP1* locus' is composed of 2.0 kb DNA fragments which contain a unique *Xba*I site, two sites for *Kpn*I, and *Sau*3A restriction enzymes. The complete nucleotide sequence has been determined. The basic repeat unit codes for two genes, one of which codes for a 246-amino acid protein, protein X, of unknown function. The smaller messenger RNA transcribed by the *CUP1* locus encodes for a 61-amino acid cysteine-rich protein, the Cu-MT.

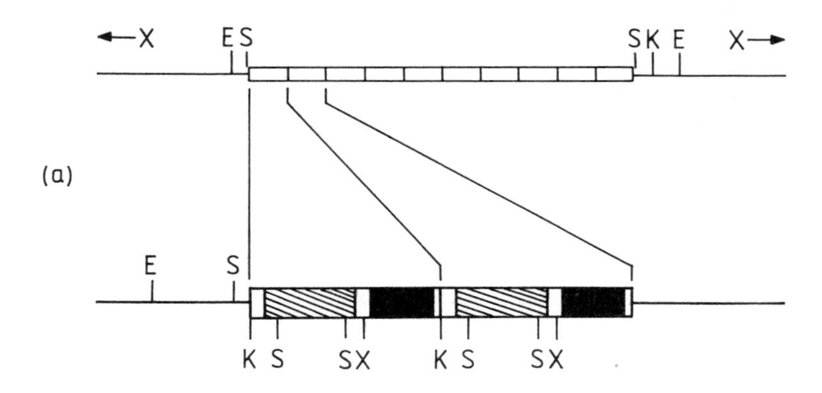

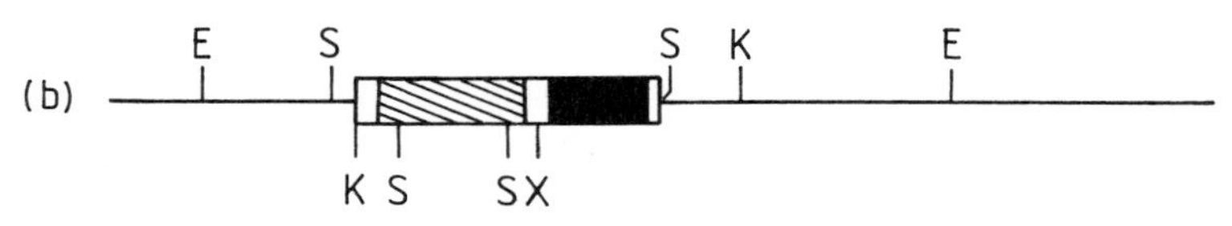

Fig. 7.4 Restriction maps of (a) CUP1^r and (b) cup1^s strains of *Saccharomyces cerevisiae*. (a) The top line represents the CUP1^r locus as ten tandemly repeated 2.0 kb units. The thin line represents the external flanking DNA sequences. Two expanded repeat units are also shown. The black region denotes the coding region of the MT gene; the shaded region is a coding region of an unknown gene. E, *Eco*RI; K, *Kpn*I; X, *Xba*I; S, *Sau*3A. *Sau*3A–*Sau*3A intervals = 0.7 and 1.3 kb; *Kpn*–*Kpn* = 2.0 kb; *Xba*–*Xba* = 2.0 kb. (b) The cup1^s locus is composed of only a single unit which is shown in expanded form. Adapted from refs 26 and 38.

Only the MT gene is transcriptionally induced after exposure to copper. Digestion of genomic DNA from a Cu-sensitive strain did not reveal a 2.0 kb DNA fragment, indicating that cupl^s strains contain a single *CUP1* locus.[26]

Although all laboratory strains of *S. cerevisiae* examined so far contain the 2.0 kb repeat unit of the *CUP1* locus and identical function sequences between the loci, industrial strains display considerable variation in the size of the *CUP1* locus. Hybridization analysis with the *CUP1* DNA does not reveal any *CUP1* sequences in these strains, yet some of the industrial strains display considerable resistance to copper.[26] In such strains, mechanisms other than amplification of the *CUP1* locus or induction of MT is involved in copper resistance such as decreased uptake.[51,126]

The basis of copper resistance in *S. cerevisiae* involves amplification of the *CUP1* locus. Gene amplification is one of the few mechanisms capable of achieving increased production of certain cellular macromolecules and this can occur under selection pressure. In simple terms, the gene amplification model of *CUP1* suggests that, due to the proposed homology of DNA sequences in the *CUP1* repeats, one or multiple units are looped out. If the loop is replicated, the copy number increases (amplification) but if the loop is degraded, the copy number decreases (deamplification). For more precise accounts of this model, consultation of more detailed works is recommended.[26,37,38]

In *S. cerevisiae*, evidence suggests that MT gene transcription may be positively regulated by a copper-specific *trans*-acting factor(s). A variety of results also suggest that the yeast MT gene may be under a negative control and there is evidence for the presence of glucose-repressible DNA sequences between the transcription initiation site and −100 bp of the MT gene.[26] A current model for yeast MT gene regulation states that autoregulation of MT synthesis is the result of copper binding by the protein which lowers the free intracellular Cu^{2+} concentration and thus dissociates the copper and copper-specific *trans*-acting transcription factor. The inactive *trans*-acting factor is activated after copper binding which restores transcriptional activity. MT gene transcription is derepressed in the presence of galactose as a sole carbon source. However, the MT gene is induced by copper even in the presence of glucose and it therefore seems that the copper-mediated activation of transcription overrides any glucose-mediated gene repression. Although yeast MT can bind a range of metals, only copper induces gene transcription.[26] Thus, cadmium does not regulate the gene although it is bound by yeast MT. The use of cadmium enables separation of gene regulation from gene function. The lack of MT induction by cadmium indicates the existence of one or more *trans*-acting factors specific for copper and also that MT protein acts as an indirect autoregulator of its own gene expression by controlling free intracellular copper.

There is also genetical analysis of metal-regulated genes from other fungi although much less is known than for *S. cerevisiae*. The gene for *N. crassa* MT protein has been cloned and the nucleotide sequence determined; this gene codes for a 26-amino acid protein.[83] The metal-regulatory mechanisms of *N. crassa* and *S. cerevisiae* appear similar but no DNA sequence homology is found in upstream regulatory sequences of the gene.[26] A strain of *Candida albicans* has been found containing DNA sequences which hybridize with *S. cerevisiae* MT and the copper-inducible proteins also show similarities with the *S. cerevisiae* MT.[26]

The value of yeast in biotechnology is considerable and well-known, and recent advances in molecular biology and genetics have confirmed the status of *S. cerevisiae* as probably the most important microbial model system for eukaryotic studies. The metal-regulated genetic system described provides a powerful tool in several areas of pure and applied significance. Metal-regulated DNA sequences appear to be the most efficient elements for heterologous gene expression discovered to date and a variety of proteins have been expressed in the *CUP1* expression system for commercial and research purposes including human serum albumin, human ubiquitin gene, *E. coli galK* gene, monkey MT-I and MT-II, mutant yeast MT and human hepatitis virus antigen.[26] The observed copper resistance is directly controlled by the copy number of the MT gene or the efficiency of MT expression and so yeast MT is valuable in yeast genetics. Areas of success include analysis of the role of centromere sequences in chromosome 3 (CEN3) and studies of proteolysis. In addition to these, brewing strains of *S. cerevisiae* are not as amenable to classical genetical studies without alteration of the desirable characteristics of the strain. The yeast MT gene has been successfully used to transform several brewing strains.[60] Selection with copper is thus a useful tool for the genetical manipulation of commercial brewing strains and improvement of the product.[26] Yeast MT may also be of potential in metal recovery since it can bind other metals besides copper, including cadmium, zinc, cobalt, silver and gold, although unlike copper these elements do not induce MT synthesis. However, yeast strains with constitutive expression of MT genes may accumulate greater amounts of metals. Alternatively, systems may be engineered for the secretion of MT proteins from yeast cells by attaching appropriate secretory signal sequences to MT genes with the related possibility of engineering a variety of MT molecules with specific affinities for different metals.

Future prospects

As described, a variety of survival strategies are exhibited by microorganisms in response to heavy metals as well as certain non-directed physico-chemical processes such as biosorption. Further understanding of such mechanisms is necessary, not only in a scientific context but also because of their potential in biotechnology. At a scientific level, the present ascendancy of molecular biology and genetics will undoubtedly contribute and some areas, notably bacterial mercury resistance and yeast metallothionein, are well advanced in this respect.[25,26,38,40,60,106] However, several areas are lacking in fundamental microbiological and biochemical studies of the physiology of resistance which is a necessary foundation for molecular analysis.[106] In addition, ecological studies are fraught with difficulty and there are the well-known problems of extrapolation of laboratory data to the field situation.[47] Despite this, environmentally related studies are likely to attain a new prominence because of the problems of accelerating environmental pollution by heavy metals and radionuclides by legal, illegal and accidental means, and also because of the new awareness of the use of extremophiles in novel biotechnological processes.

Both living and dead microbial biomass can accumulate heavy metals and some processes are already in commercial operation.[46] It should be stressed that microbe-based processes for metal removal/recovery may not necessarily replace conventional technologies but may serve as 'polishers' to processes not completely efficient. Selectivity is a problem for many microbial systems and this is worthy of further attention in the future. Some degree of selectivity in both biosorption and desorption may be possible using strain selection and/or manipulation of external conditions. Living cells do have possibilities, though internalized metal may be difficult to recover and toxicity may be a problem. Although resistant strains can be readily isolated or engineered, they often exhibit reduced accumulation. However, there are examples where resistant yeast strains take up more metal than sensitive strains, probably by means of more efficient internal sequestration, and where resistant microbial consortia are also highly effective. Currently systems that use living cells appear relatively unsophisticated, and metallothioneins, particulate metal accumulation, siderophores and analogous molecules and extracellular precipitation have all received scant exploitation. With the realization that yeast metallothionein and siderophores may be able to bind precious metals and the possibility of genetic manipulation of organisms to, for example, constitutive metallothionein expression, this may change in the future.

In relation to industrial applications of microbial metal removal/recovery, most attention has so far focused on high-value elements of commercial importance. With non-precious metals and radionuclides, the economics are frequently unattractive and there is the problem of contaminated biomass. However, the containment of low-volume waste or eluate is preferable to its entry into the environment. At the moment, environmental protection receives only limited political and industrial attention. However, with increasing pollution and public awareness of this and the dangers of heavy metals and radionuclides to all living components of the biosphere, action may be taken and the fuller involvement of microbe-based technologies may be part of this. Unfortunately, for all those concerned with environmental safety, optimism is rather a scarce commodity!

Conclusions

Toxic heavy metals and radionuclides have adverse effects on aquatic and terrestrial ecosystems. Despite this, microorganisms are commonly found in polluted habitats and possess a range of morphological and physiological attributes that enable survival. Certain kinds of resistances may be genetically determined and transferable whereas, in other cases, tolerance may depend on intrinsic properties of the organism and/or detoxification due to environmental components. Many organisms can express a variety of survival strategies and, as well as providing excellent model systems for scientific endeavour in fields ranging from ecology to molecular biology and genetics, are also of biotechnological relevance to the detoxification of industrial effluents, metal recovery, land reclamation, the preservation of natural and synthetic materials and treatment of infection.

References

1. Adjimani, J.P. and Emery, T. (1987) *J. Bacteriol.*, **169**, 3664–8.
2. Aiking, H., Govers, H. and van't Riet, J. (1985) *Appl. Environ. Microbiol.*, **50**, 1262–7.
3. Al-Atia, G.R. (1980) *J. Protozool.*, **27**, 128–32.
4. Ashida, J. (1965) *Ann. Rev. Phytopathol.*, **3**, 153–74.
5. Babich, H. and Stotzky, G. (1980) *CRC Crit. Rev. Microbiol.*, **8**, 99–145.
6. Babich, H. and Stotzky, G. (1985) *Environ. Res.*, **36**, 11–137.
7. Babich, H. and Stotzky, G. (1987) *Adv. Appl. Microbiol.*, **23**, 55–117.
8. Bedell, G.W. and Soll, D.R. (1979) *Infect. Immun.*, **26**, 348–54.
9. Bell, A.A. and Wheeler, M.H. (1986) *Ann. Rev. Phytopathol.*, **24**, 411–51.
10. Belliveau, B.H., Starodub, M.E., Cotter, C. and Trevors, J.T. (1987) *Biotech. Adv.*, **5**, 101–27.
11. Belly, R.T. and Kydd, G.C. (1982) *Dev. Ind. Microbiol.*, **23**, 567–77.

12. Beveridge, T.J. and Murray, R.G.E. (1980) *J. Bacteriol.*, **141**, 876–87.
13. Beveridge, T.J., Meloche, J.D., Fyfe, W.S. and Murray, R.G.E. (1983) *Appl. Environ. Microbiol.*, **45**, 1094–108.
14. Bewley, R.J.F. and Campbell, R. (1980) *Microbial Ecol.*, **6**, 227–40.
15. Bewley, R.J.F. and Stotzky, G. (1983) *Sci. Total Environment*, **31**, 41–55.
16. Bewley, R.J.F. and Stotzky, G. (1983) *Sci. Total Envionment*, **31**, 57–69.
17. Bitton, G. and Freihofer, V. (1978) *Microbial Ecol.*, **4**, 119–25.
18. Bonhomme, A., Quintana, C. and Duraud, M. (1980) *J. Protozool.*, **27**, 491–7.
19. Borst-Pauwels, G.W.F.H. (1981) *Biochim. Biophys. Acta*, **650**, 88–127.
20. Bradley, R., Burt, A.J. and Read, D.J. (1982) *New Phytol.*, **91**, 197–209.
21. Brierley, J.A. and Brierley, C.L. (1983), in P. Westbroek and E.W. de Jong (eds) *Biomineralization and Biological Metal Accumulation*, Dordrecht, Reidel, pp. 499–509.
22. Brierley, C.L., Kelly, D.P., Seal, K.J. and Best, D.J. (1985), in I.J. Higgins, D.J. Best and J. Jones (eds) *Biotechnology*, Oxford, Blackwell, pp. 163–212.
23. Brierley, J.A., Goyak, G.M. and Brierley, C.L. (1986), in H. Eccles and S. Hunt (eds) *Immobilisation of Ions by Bio-sorption*, Chichester, Ellis Horwood, pp. 105–17.
24. Brock, T.D. (1987), in M. Fletcher, T.R.G. Gray and J.G. Jones (eds) *Ecology of Microbial Communities*, Cambridge University Press, pp. 1–17.
25. Brown, N.L. (1985) *Trends Biochem. Sci.*, **10**, 400–3.
26. Butt, T.R. and Ecker, D.J. (1987) *Microbiol. Rev.*, **51**, 351–64.
27. Butt, T.R., Sternberg, E.J., Gorman, J.A., Clark, P., Hamer, D., Rosenberg, M. and Crooke, S.T. (1984) *Proc. Nat. Acad. Sci. USA*, **81**, 3332–6.
28. Charley, R.C. and Bull, A.T. (1979) *Arch. Microbiol.*, **123**, 239–44.
29. Clarke, S.E., Stuart, J. and Sanders-Loehr, J. (1987) *Appl. Environ. Microbiol.*, **53**, 917–22.
30. Cooney, J.J. (1988), in P.J. Craig and F. Glockling (eds) *The Biological Alkylation of Heavy Elements*, London, Royal Society of Chemistry, pp. 92–104.
31. Darnall, D.W., Greene, B., Henzl, M.J., Hosea, J.M., McPherson, R.A., Sneddon, J. and Alexander, M.D. (1986) *Environ. Sci. Technol.*, **20**, 206–8.
32. Dekker, J. (1976) *Ann. Rev. Phytopathol.*, **14**, 405–28.
33. Dunn, G.M. and Bull, A.T. (1983) *Eur. J. Appl. Microbiol. Biotechnol.*, **17**, 30–4.
34. Duxbury, T. (1985), in K.C. Marshall (ed.) *Advances in Microbial Ecology*, New York, Plenum Press, pp. 185–235.
35. Duxbury, T. (1986) *Microbiol. Sci.*, **3**, 330–3.
36. Duxbury, T. and Bicknell, B. (1983) *Soil Biol. Biochem.*, **15**, 243–50.
37. Fogel, S. and Welch, J.W. (1982) *Proc. Nat. Acad. Sci. USA*, **79**, 5342–6.
38. Fogel, S., Welch, J.W. and Karin, M. (1983) *Curr. Genetics*, **7**, 1–9.
39. Foster, T.J. (1983) *Microbiol. Rev.*, **47**, 361–409.
40. Foster, T.J. (1987) *CRC Crit. Rev. Microbiol.*, **15**, 117–40.
41. Foye, W.O. (1977), in E.D. Weinberg (ed.), *Microorganisms and Minerals*, New York, Marcel Dekker, pp. 387–419.
42. Freedman, B. and Hutchinson, T.C. (1980) *Can. J. Bot.*, **58**, 1722–36.
43. Gadd, G.M. (1986), in B.J. Dutka and G. Bitton (eds) *Toxicity Testing using Microorganisms*, Boca Raton, CRC Press, pp. 43–77.
44. Gadd, G.M. (1986), in H. Eccles and S. Hunt (eds) *Immobilisation of Ions by Bio-sorption*, Chichester, Ellis Horwood, pp.135–47.
45. Gadd, G.M. (1986), in R.A. Herbert and G.A. Codd (eds) *Microbes in Extreme Environments*, London, Academic Press, pp. 83–110.
46. Gadd, G.M. (1989), in H.-J. Rehm and G. Reed (eds) *Biotechnology — A Comprehensive Treatise, Vol. 6b*, Weinheim, VCH Verlagsgesellschaft, in press.

47. Gadd, G.M. and Griffiths, A.J. (1978) *Microbial Ecol.*, **4**, 303–17.
48. Gadd, G.M. and White, C. (1985) *J. Gen. Microbiol.*, **131**, 1875–9.
49. Gadd, G.M. and White, C. (1989), in G.M. Gadd and R.K. Poole (eds) *Metal–Microbe Interactions*, Oxford, IRL Press.
50. Gadd, G.M., Chudek, J.A., Foster, R. and Reed, R.H. (1984) *J. Gen. Microbiol.*, **130**, 1969–75.
51. Gadd, G.M., Stewart, A., White, C. and Mowll, J.L. (1984) *FEMS Microbiol. Lett.*, **24**, 231–4.
52. Gadd, G.M., Mowll, J.L., White, C. and Newby, P.J. (1986) *Tox. Assess.*, **1**, 169–85.
53. Gadd, G.M., White, C. and Mowll, J.L. (1987) *FEMS Microbiol. Ecol.*, **45**, 261–7.
54. Gadd, G.W., White, C. and de Rome, L. (1988), in P.R. Norris and D.P. Kelly (eds) *Biohydrometallurgy*, Kew, Science and Technology Letters, pp. 421–35.
55. Gale, N.L. and Wixson, B.G. (1979) *Dev. Ind. Microbiol.*, **20**, 259–73.
56. Grill, E., Winnaker, E.L. and Zeuk, M.H. (1986) *FEBS Lett.*, **197**, 115–20.
57. Hallas, L.E. and Cooney, J.J. (1981) *Appl. Environ. Microbiol.*, **41**, 446–71.
58. Hansen, C.L., Zwolinski, G., Martin, D. and Williams, J.W. (1984) *Biotechnol. Bioeng.*, **26**, 1330–3.
59. Hausinger, R.P. (1987) *Microbiol. Rev.*, **51**, 22–42.
60. Henderson, R.C.A., Cox, B.S. and Tubb, R. (1985) *Curr. Genetics*, **9**, 133–8.
61. Heuillet, E., Guerbette, F., Genou, C. and Kader, J.C. (1988) *Int. J. Biochem.*, **20**, 203–10.
62. Higham, D.P., Sadler, P.J. and Scawen, M.D. (1986) *J. Gen. Microbiol.*, **132**, 1472–82.
63. Hollo, J., Toth, J., Tengerdy, R.P. and Johnson, J.E. (1979, in K. Venkatsubramanian (ed.) *Immobilized Microbial Cells*, Washington, DC, American Chemical Society, pp. 73–86.
64. Hutchins, S.R., Davidson, M.S., Brierley, J.A. and Brierley, C.L. (1986) *Ann. Rev. Microbiol.*, **40**, 311–36.
65. Jeffries, T.W. (1982), in M.J. Bull (ed.) *Progress in Industrial Microbiology*, Amsterdam, Elsevier, pp. 21–75.
66. Jensen, T.E., Baxter, M., Rauchlin, J.W. and Jani, V. (1982) *Environ. Pollution*, **27**, 119–27.
67. Jernelov, A. (1973), in M.J. Chadwick and G.T. Goodman (eds) *The Ecology of Resource Degradation and Renewal*, Oxford, Blackwell, pp. 49–55.
68. Joho, M., Yamanaka, C. and Murrayama, T. (1986) *Microbios*, **45**, 169–79.
69. Jordan, M.J. and Lechevalier, M.P. (1975) *Can. J. Microbiol.*, **21**, 1855–65.
70. Kelly, D.P., Norris, P.R. and Brierley, C.L. (1979), in A.T. Bull, D.C. Ellwood and C. Ratledge (eds) *Microbial Technology, Current State, Future Prospects*, Cambridge University Press, pp. 263–308.
71. Kessels, B.G.F., Belde, P.J.M. and Borst-Pauwels, G.W.F.J. (1985) *J. Gen. Microbiol.*, **131**, 1533–7.
72. Kiff, R.J. and Little, D.R. (1986), in H. Eccles and S. Hunt (eds) *Immobilisation of Ions by Biosorption*, Chichester, Ellis Horwood, pp. 71–80.
73. Kogut, M. (1980) *Trends Biochem. Sci.*, **5**, 15–17.
74. Kogut, M. (1980) *Trends Biochem. Sci.*, **5**, 47–50.
75. Lester, J.N., Sterritt, R.M., Rudd, T. and Brown, M.J. (1984), in J.M. Grainger and J.M. Lynch (eds) *Microbiological Methods for Environmental Biotechnology*, London, Academic Press, pp. 197–217.
76. Macaskie, L.E. and Dean, A.C.R. (1984) *J. Gen. Microbiol.*, **130**, 53–62.
77. Macaskie, L.E. and Dean, A.C.R. (1987) *Enzyme Microb. Technol.*, **9**, 2–4.

78. Macaskie, L.E., Dean, A.C.R., Cheetham, A.K., Jakeman, R.J.B. and Skarnulis, J. (1987) *J. Gen. Microbiol.*, **133**, 539–44.
79. Macaskie, L.E., Wates, J.M. and Dean, A.C.R. (1987) *Biotechnol. Bioeng.*, **30**, 66–73.
80. Mitra, R.S. (1984) *Appl. Environ. Microbiol.*, **47**, 1012–16.
81. Moore, J.M. (1981) *Wat. Res.*, **15**, 97–105.
82. Mowll, J.L. and Gadd, G.M. (1985) *Trans. Br. Mycol. Soc.*, **84**, 684–9.
83. Munger, K., Germann, N.A. and Lerch, K. (1985) *EMBO J.*, **4**, 1459–62.
84. Murphy, R.J. and Levy, J.F. (1983) *Trans. Br. Mycol. Soc.*, **81**, 165–8.
85. Nakajima, A. and Sakaguchi, T. (1986) *Appl. Microbiol. Biotechnol.*, **24**, 59–64.
86. Nakamura, Y., Katayama, S., Okada, Y., Suzuki, F. and Nagata, Y. (1981) *Agric. Biol. Chem.*, **45**, 1167–72.
87. Newby, P.J. and Gadd, G.M. (1988), in P.J. Craig and F. Glockling (eds) *The Biological Alkylation of Heavy Elements*, London, Royal Society of Chemistry, pp. 164–7.
88. Nieboer, E. and Richardson, D.H.S. (1980) *Environ. Pollution*, **1**, 3–26.
89. Nielands, J.B. (1981) *Ann. Rev. Biochem.*, **50**, 715–31.
90. Norberg, A.B. and Rydin, S. (1984) *Biotechnol. Bioeng.*, **26**, 265–8.
91. Nordgren, A., Baath, E. and Soderstrom, B. (1983) *Appl. Environ. Microbiol.*, **45**, 1829–37.
92. Ochiai, E.I. (1987) *General Principles of Biochemistry of the Elements*, New York, Plenum Press.
93. Okorokov, L.A. (1985), in I.S. Kulaev, E.A. Dawes and D.W. Tempest (eds) *Environmental Regulation of Microbial Metabolism*, London, Academic Press, pp. 339–49.
94. Olafson, R., Abel, W.K. and Sim, R.G. (1979) *Biochem. Biophys. Res. Comm.*, **89**, 36–43.
95. Pooley, F.E. (1982) *Nature*, **296**, 642–3.
96. Rai, L.C., Gaur, J.P. and Kumar, H.D. (1981) *Biol. Rev.*, **56**, 99–151.
97. Raymond, K.N., Muller, G. and Matzanke, B.F. (1984) *Topics Curr. Chem.*, **123**, 49–102.
98. Robinson, J.B. and Tuovinen, O.H. (1984) *Microbiol. Rev.*, **48**, 95–124.
99. Ross, I.S. (1975) *Trans. Br. Mycol. Soc.*, **64**, 175–93.
100. Ruthven, J.A. and Cairns, J. (1973) *J. Protozool.*, **20**, 127–35.
101. Shumate, S.E. and Strandberg, G.W. (1985), in M. Moo-Young, C.N. Robinson and J.A. Howell (eds) *Comprehensive Biotechnology, Vol. 4*, New York, Pergamon Press, pp. 235–47.
102. Shumate, S.E., Strandberg, G.W., McWhirter, D.A., Parrott, J.R., Bogacki, G.M. and Locke, B.R. (1980) *Biotechnol. Bioeng. Symp.*, **10**, 27–34.
103. Silver, S. (1978), in B.P. Rosen (ed.) *Bacterial Transport*, New York, Marcel Dekker, pp. 221–324.
104. Silver, S. (1983), in P. Westbroek and E.W. de Jong (eds) *Biomineralization and Biological Metal Accumulation*, Dordrecht, Reidel, pp. 439–57.
105. Silver, S. (1984), in J.O. Nriagu (ed.) *Changing Metal Cycles and Human Health*, Berlin, Springer-Verlag, pp. 199–223.
106. Silver, S. and Misra, T.K. (1988), in P.J. Craig and F. Glockling (eds) *The Biological Alkylation of Heavy Elements*, London, Royal Society of Chemistry, pp. 211–42.
107. Spisak, F. (1979) *Dev. Ind. Microbiol.*, **20**, 249–57.
108. Sterritt, R.M. and Lester, J.N. (1980) *Sci. Total Environment*, **14**, 5–17.
109. Sterritt, R.M. and Lester, J.N. (1986), in H. Eccles and S. Hunt (eds) *Immobilisation of Ions by Bio-sorption*, Chichester, Ellis Horwood, pp. 121–34.
110. Strandberg, G.W., Shumate, S.E. and Parrott, J.R. (1981) *Appl. Environ. Microbiol.*, **41**, 237–45.
111. Summers, A.O. and Silver, S. (1978) *Ann. Rev. Microbiol.*, **32**, 637–72.

112. Surowitz, K.G., Titus, J.A. and Pfister, M. (1984) *Arch. Microbiol.*, **140**, 107–12.
113. Townsley, C.C., Ross, I.S. and Atkins, A.S. (1986), in H. Eccles and S. Hunt (eds) *Immobilisation of Ions by Bio-sorption*, Chichester, Ellis Horwood, pp. 159–70.
114. Trevors, J.T. (1986) *J. Basic Microbiol.*, **26**, 499–504.
115. Trevors, J.T. (1987) *Enzyme Microb. Technol.*, **9**, 331–3.
116. Trevors, J.T. (1987) *Microbiol. Sci.*, **4**, 29–31.
117. Trevors, J.T., Oddie, K.M. and Belliveau, B.H. (1985) *FEMS Microbiol. Rev.*, **32**, 39–54.
118. Trevors, J.T., Stratton, G.W. and Gadd, G.M. (1986) *Can. J. Microbiol.*, **32**, 447–64.
119. Tsezos, M. (1984) *Biotechnol. Bioeng.*, **26**, 973–81.
120. Tsezos, M. (1986), in H. Eccles and S. Hunt (eds) *Immobilisation of Ions by Bio-sorption*, Chichester, Ellis Horwood, pp. 201–18.
121. Tuovinen, O.H. and Kelly, D.P. (1974) *Int. Metall. Rev.*, **19**, 21–31.
122. Vaccaro, R.F., Azam, F. and Hodson, R.E. (1977) *Bull. Mar. Sci.*, **27**, 17–22.
123. Vymazal, J. (1987) *Tox. Assess.*, **2**, 387–415.
124. Wainwright, M., Grayston, S.J. and de Jong, P. (1986) *Enzyme Microb. Technol.*, **8**, 597–600.
125. Welch, J.W., Fogel, S., Cathala, G. and Karin, M. (1983) *Mol. Cell. Biol.*, **8**, 1353–61.
126. White, C. and Gadd, G.M. (1986) *FEMS Microbiol. Ecol.*, **38**, 277–83.
127. White, C. and Gadd, G.M. (1987) *J. Gen. Microbiol.*, **133**, 727–37.
128. Williams, J.I. and Pugh, G.J.F. (1975) *Trans. Br. Mycol. Soc.*, **64**, 255–63.
129. Williams, J.W. and Silver, S. (1984) *Enzyme Microb. Technol.*, **6**, 530–7.
130. Wood, J.M. and Wang, H.K. (1983) *Environ. Sci. Technol.*, **17**, 582–90.
131. Yakubu, N.A. and Dudeney, A.W.L. (1986), in H. Eccles and S. Hunt (eds) *Immobilisation of Ions by Bio-sorption*, Chichester, Ellis Horwood, pp. 183–200.

Species index

Subject index